Understanding the financial crisis:
investment, risk and governance

Understanding the financial crisis:
investment, risk and governance

First edition 2009

ISBN 978-87-993151-0-9

Printed in Denmark by Nofoprint as

SimCorp StrategyLab

This volume is a result of the research programme of SimCorp StrategyLab, an independent research institution founded and managed by SimCorp.

The research work of SimCorp StrategyLab focuses on identifying, understanding and suggesting solutions to issues pertaining to mitigating risk, reducing cost and enabling growth in the investment management industry.

The contributions to this anthology do not necessarily reflect the views of SimCorp StrategyLab. SimCorp StrategyLab has used all reasonable endeavours to ensure the accuracy of the information, however, SimCorp StrategyLab does not guarantee or warrant the accuracy or completeness, factual correctness or reliability of any information in this volume and does not accept any liability for any errors or omissions including any inaccuracies or typographical errors.

SimCorp StrategyLab
c/o SimCorp A/S
Weidekampsgade 16
2300 Copenhagen S
Denmark

www.simcorpstrategylab.com

Steen Thomsen, Caspar Rose
and Ole Risager (eds.)

Understanding the financial crisis: investment, risk and governance

Contents

List of figures and tables

Foreword

The world has witnessed a global, financial pandemic; the worst financial crisis since the Great Depression. Many financial institutions that were commonly perceived as too big to fail did exactly that. The high levels of risk taking combined with what in some cases looked like greed appears to be driven by harmful social constructivism, to a large degree unchallenged by regulators, governance, risk management functions, systems and models. As a result, many people worldwide have lost their homes, their jobs, their savings and not least their confidence in the financial sector and the laws and entities regulating it.

This volume is the first in a series of three, published by SimCorp StrategyLab, resulting from its research programmes within the investment management industry, focusing on mitigating risk, reducing cost and enabling growth. The book gives suggestions as to why the financial crisis happened. It presents thoughts, proposals and recommendations on how to avoid similar, future meltdowns in the financial sector and the financial markets and discusses the severity of the resulting negative impact on the real economy which the world has been seeing.

The book is an anthology. It reflects experience, research and thought leadership from experts and specialists in various disciplines: finance, economics, mathematics, risk, governance, compliance, IT and regulation among others, serving as a contribution to the work-in-progress aimed at building a new, sustainable financial system. The authors are for the most part senior scholars of respected research institutions; including INSEAD; Stern School of Business, New York University; University of Toronto; UQ Business School, University of Queensland and Copenhagen Business School.

At the same time this work is being accompanied by thoughts and recommendations from highly experienced executives at major firms in the financial sector, such as Deutsche Bank, ATP and Allianz SE. Last, but not least, the crisis in the financial industry and its implications for individual institutions are viewed from a strategic IT architectural perspective by specialists in financial IT architecture at SimCorp.

On behalf of SimCorp StrategyLab, I would like to thank all the contributors for the efforts they have made to the making of this book.

Throughout this volume it has been our aim to contribute to the identification and explanation of the drivers behind the crisis in the financial industry. Moreover, it has also been our intention to present thoughts on how to move forward. The book is written with management of the global investment management industry in mind. Yet, policy-makers, legislators, financial journalists, students, and everyone else who has a stake or interest in the global financial industry *per se* will find the contents highly relevant for their understanding of the past as well as for the decisions they will have to take in the future.

Lars Bjørn Falkenberg
Assistant Director of SimCorp StrategyLab

INTRODUCTION

1 Macroeconomic perspectives on the financial crisis: an overview

by Ole Risager

Ole Risager, Ph.D, is a professor at Copenhagen Business School. He has published extensively on foreign exchange markets, stock markets, and macroeconomics. His work has appeared in international journals as well as in publications of the International Monetary Fund (IMF) and the World Bank. He has previously served as a senior economist to the IMF in Washington and as a consultant to the World Bank. He has also been Vice President and Chief Economist at the Danish shipping and oil company A.P. Moller – Maersk. Ole Risager is currently chairman for Core German Residential II, a real estate investment company, supervisory board member of LD Invest, a mutual investment fund, and supervisory board member of H+H, a Danish manufacturing company. Contact: or.int@cbs.dk

This chapter deals with the macroeconomic roots of the financial crisis. We argue that there are several important macroeconomic factors that together with excessive risk-taking in the financial sector, reflecting poor corporate governance and to some extent also inadequate regulation and enforcement, led to the current crisis. Once the crisis had hit the system, high leverage is the key to understanding why the crisis turned into the worst financial crisis since the Great Depression.

At the end of the chapter, we discuss whether we can avoid crises in the future through improvements in policy-making. We argue that crises are inevitable and that we can never fully avoid them. That said, it should be noted that there has been progress in crisis prevention. The best example of this is that the world avoided a new depression due to timely and effective global policy intervention led by the US Federal Reserve and the Treasury. Had policy-makers not intervened and undertaken the necessary fire fighting the financial crisis could very well have led the world into a very deep and prolonged recession reflecting that world industrial production, trade, and stock markets were diving faster from April 2008 to early spring 2009 than during 1929–30, see Eichengreen and O'Rourke (2009).[1]

As a result of these interventions, the US government had to take over some of the

---

1   April 2008 marks the peak of world industrial production.

world's largest financial institutions. Similarly, the financial sector has also to some extent become state-owned in other countries.[2] The price of this insurance against a meltdown of leading financial institutions and the world economy is accelerating public sector deficits in all advanced economies. Taxpayers will have to carry this enormous burden in the coming years.[3]

From a business point of view the lesson is once again that it pays to have a strong focus on risk management including macroeconomic surveillance. Companies that are forward looking and know that the economic cycle is a fundamental characteristic of the world we live in can better hedge and take advantage of the cycle than companies that disregard macroeconomic realities, see Risager (2009).

Macroeconomic roots of the financial crisis

The roots of the financial crisis go back to the preceding long period of low interest rates, high growth, excessive optimism and risk willingness, and high leverage spurred by both the boom itself and by the creation of extensive shadow banking and new financial products. These meant that the financial sector's capital reserves were inadequate in a crisis situation. Below we discuss the different macroeconomic factors that played a role in the build-up of the crisis. We also discuss estimates of likely losses and write-downs due to the IMF (2009a). If these estimates prove to be correct we should also expect large losses going forward. Moreover, the latest round of capital injections into the US banking sector in May 2009, following the US Treasury's stress tests of major US financial institutions, may prove to be inadequate.

Low interest rates and the boom-bust global housing cycle

In response to the 2001 recession, caused by the bursting of the stock market bubble and the sharp cutback in firms' capital spending, the US Federal Reserve reduced the Fed funds rate to 1 percent, down from a peak at 5.5 percent prior to the recession. The Fed's aggressive interest rate policy led a global monetary policy easing trend reflecting that other countries were also hit by the downturn in the US. Moreover, as the world economy operated with subdued inflation pressure, reflecting the integration of China and other low-cost producers into the global supply chain and reflecting reduced trade union bargaining power, central banks could maintain this low inter-

---

2   The US government now owns the world's largest mortgage companies Fannie Mae and Freddy Mac, the world's largest insurance company AIG, and is an important shareholder of major US banks. Similarly, the UK government is now a majority shareholder of some of the largest financial institutions in the UK, including RBS and Lloyds TSB; only Barclays Bank and HSBC managed to run their own affairs.

3   The public deficits are right now absorbed by central banks but they cannot in the long-term finance the deficits assuming that they will not violate their inflation targets. That said, one can of course not rule out that the crisis in the medium-to-long term will lead to higher inflation and to a breakdown of inflation targeting regimes.

est rate policy over an extended period. This also mirrored the dominant view at the time that central banks should only worry about inflation of the prices of goods and services. Central banks should not be concerned about asset price inflation including the booming housing sector. It is easier to clean up after a bubble has run its course than to identify and prick a bubble as the Fed Chairman Alan Greenspan put it. This view was not shared by all economists. Roubini (2008) warned against a financial pandemic and there were also others who saw a financial crisis in the horizon.[4,5]

There is little doubt that this long period of cheap global credit is an important explanation of the sharp run up in house prices we have seen in almost all advanced economies with the exception of Japan and Germany, see also Buiter (2008).[6] On top of this, mortgage institutions and banks on a global scale undertook a number of innovations that essentially made it possible to buy real estate with very little down payment, if any at all. The problems with some of the innovations in mortgage financing became particular obvious in the US sub-prime market in late 2006 and early 2007, see Shiller (2008). The sub-prime loans were often made in anticipation that house prices would continue to rise. In many cases sub-prime mortgage loans were so-called 'teaser loans' with low repayments, if any, at the beginning and with low variable interest rates taking advantage of the low rates at the short end of the interest rate yield curve. The idea was basically that even though many of the new homeowners could not afford a house, rising real estate prices would eventually solve that problem. However, when house prices flattened out in autumn 2006 and later nosedived, the problems with this high risk model cascaded and this marked the beginning of the global credit crisis. At that time sub-prime had become an important source of mortgage financing accounting for more than 20 percent of annual mortgage financing in the US.

Declining house prices and higher mortgage rates, reflecting that the boom had lasted several years, in combination with irresponsible credit scoring eventually triggered a sharp increase in delinquency rates.[7] In some cases they rose significantly above 20 percent, see figure 2. Investors who had purchased these mortgage loans, often securitised, that is, sliced and packaged together with other credit instruments,

---

4   When reading Roubini's paper, one is struck by the similarities of his predictions and what has actually happened. That said, one of the problems with some of these warnings was the timing, that is, the warnings were issued at a very early date and years before the crisis materialised.

5   Amongst theoretical economists, Minsky was the one who most frequently wrote about financial crises, see Minsky (1977). His view is basically that investors in good times take on too much debt to finance aggressive asset expansions. Eventually, investors reach a point in time where the cash generated by their assets is insufficient to service the mountains of debt. This forces investors to sell assets, which leads to plummeting asset prices, which in turn leads to even more demand for cash, and therefore to a further decline in asset prices, and so forth. When the economy has arrived at this point, we are at the so-called Minsky moment.

6   All econometric studies of house prices point to a large role of (nominal) interest rates as houses in many cases are purchased with an eye on current after-tax nominal mortgage costs.

7   These loans were also known as liars' loans because income statements were often incorrect.

8   These complicated asset backed securities were rated by rating agencies, but apparently with a flawed risk assessment.

House price index   Savings rate

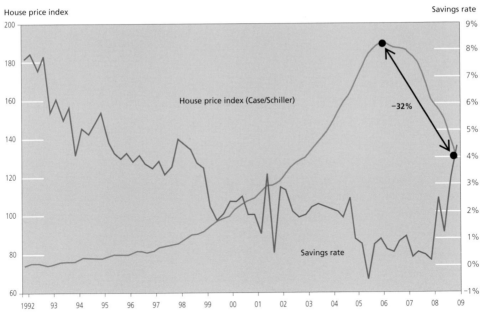

Fig. 1. US house prices and savings rate. Source: Case-Shiller and US Treasury.

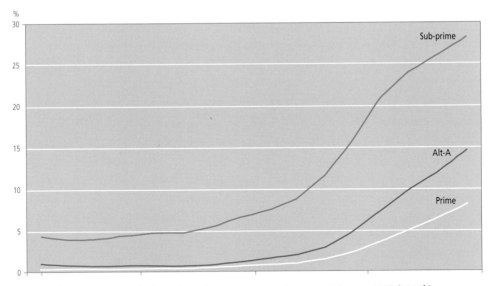

Fig. 2. Delinquency rate of US residential mortgage loans in percent. Source: IMF (2009b).

started to report large losses.[8] These investors were US and European institutions in-cluding investment and retail banks, insurance companies, pension funds, and even local governments and state-owned entities reported large losses.[9] As a result of the

---

9   Interestingly, the major sources of the financing of the US trade deficit including China, Japan and OPEC countries were only minor victims.

large losses in the banking sector, credit started to dry up and several top US investment banks, including Merrill Lynch and Bear Stearns, came under severe pressure
and were close to collapsing, had they not been taken over by Bank of America and
JPMorgan Chase, respectively.[10]

However, the prime mortgage market also suffered in response to declining US
house prices with Fannie Mae and Freddy Mac taking huge losses on their books. This
explains why the world's largest mortgage institutions, endowed with too little equity
capital, had to be rescued by the US Treasury. Like the sub-prime market, the prime
mortgage market continues to be under stress. Thus, at the end of the first quarter of
2009, the delinquency rate for all US mortgages is at a record high of 9.1 percent. This
shows that losses on mortgage assets will be significant also going forward.

It was not only housing related institutions and investment banks that were under
water. AIG, the world's largest insurance company, which had extended its business
into derivatives and other complex risk products, was rescued in September 2008 to
avoid larger, systemic consequences. Large financial sector losses and outright collapses of the world's major financial institutions meant that the interbank market
essentially dried up with interbank rates soaring. The decision of the US authorities
not to rescue Lehman Brothers and the subsequent collapse in September 2008 was
a further blow to markets. The Lehman collapse led to a sharp decline in confidence
and to a huge increase in perceived counterparty risks. Moreover, as other countries
including the UK, Germany, France, Australia, Sweden, Denmark and so forth, were
having their own financial sector crises, due to losses on domestic housing related
credits and other risky assets, we had a full blown international financial crisis.

As a result of a near collapse of financial markets, advanced economies went into
the deepest recession since the 1930s. In the fourth quarter of 2008, GDP fell by 6.3
percent in the US, 6.2 percent in the Eurozone, and 14.4 percent in Japan.[11] In the first
quarter of 2009, GDP declined by 5.7 percent in the US, 9.8 percent in the Eurozone,
and by 15.2 percent in Japan. Since then, business and other confidence indicators
point to a fall in the rate of contraction of advanced economies. Some stabilisation is
therefore around the corner and some expect a weak recovery in the second half of
2009, though the degree of visibility is unusually low.

With the benefit of hindsight it is clear that the aggressive and highly accommodative monetary policy and the unhindered development of easier mortgage finance
fuelled a new bubble, namely, a bubble in the housing market on a global scale.[12] Moreover, it also seems clear now that central banks should set policy interest rates in view
of a broader set of inflation targets including house and equity price inflation; see also
IMF (2009a). That said, it should be noted that this is a highly complex issue as there

---

10   These deals were orchestrated by the Fed.
11   Annualised rates.
12   The accommodative monetary policy strategy was frequently coined a 'Greenspan-put' in markets,
     indicating that if something went wrong in the real economy, Chairman Greenspan would cut rates and
     hence provide insurance as to how much markets could fall.

is no general recipe for how to conduct monetary policy targeting both goods and asset price inflation. Thus, there is no easy-to-use definition of asset price inflation and there is no formula for how central banks should strike a balance between asset price inflation and goods price inflation.

Excessive optimism and risk taking in other sectors and regions

Excessive optimism and risk taking spurred by the boom itself is also a feature outside the banking and real estate sector. During this cycle the emergence of China and India has in particular spurred optimism. The tremendous outsourcing opportunities and the fundamental change of the world's supply chains have led to massive investments in the global shipping industry in a way that very much resembles the pig-cycle known from agriculture.[13] As a result of this, container shipping, tank and bulk are all stuck with massive excess capacity and plummeting rates. It is likely that it will take years before capacity is in line with demand and years before these industries again post a reasonable return on assets.

At the regional level this cycle has witnessed a remarkable building boom in Dubai and other oil rich countries driven by a high oil price and the desire to diversify these countries into tourism and banking.[14] As a result of this policy-generated hype, real estate prices skyrocketed but are now down by more than 50 percent, whereas vacancy rates are soaring.

Excessive risk taking is also evident in corporate finance. Many companies have used their cash flows to reduce their share capital, which boosted earnings per share and hence made companies look attractive by conventional price multiples. However, the corporate sector is now globally reversing this process, that is, firms are now raising equity capital to make the balance sheet look more acceptable to banks, and in some cases firms also use the funding to pay back banks and other creditors. Similarly, capital funds, often gearing their investments tremendously relative to historic norms, now also live a more quiet life. Strong balance sheets are fashionable again and 'cash is king'.

IMF's estimates of total losses over the period 2007–10

The International Monetary Fund (IMF) has estimated the likely total losses on assets originating in the US, the Eurozone including the UK, and Japan. A key assumption is that the US will first start a weak recovery in 2010. The Eurozone and the UK will lag one year behind, that is, GDP will also fall in 2010, whilst Japan will track US developments, see IMF (2009b) for further details.

---

13 High prices trigger large investments of many companies at one and the same time. So what was a smart idea for a single firm turns out to be a disaster for the whole industry since capacity is expanded far beyond what is needed. The result is a sharp decline in prices and bankruptcies of many individual companies. It is through this process the industry eventually reaches equilibrium.

14 In this cycle oil peaked at $147 per barrel.

IMF estimates of potential write-downs by origin of assets ($ billion)				
United States		**Banks**	**Insurers**	**Other***
Loans	1068	601	53	414
Securities	1644	1002	164	477
Total	2712	1604	218	890
Eurozone + UK		**Banks**	**Insurers**	**Other**
Loans	888	551	44	292
Securities	305	186	31	89
Total	1193	737	75	381
Japan		**Banks**	**Insurers**	**Other**
Loans	131	118	7	7
Securities	17	11	2	5
Total	149	129	8	12
Total for all loans and securities	4054			

Table 1. IMF estimates of financial sector potential write-downs, 2007–10 by geographic origin of assets as of April 2009 insurers.
* Include US government-sponsored enterprises, hedge funds, pensions, and other non-bank financial institutions. Source: IMF (2009b).

Under these assumptions, the Fund projects total losses for all loans and securities, such as most of the sub-prime bonds, originated in the US, to be about $2,712 billion.[15] As of today we have seen less than half of these realised. If these estimates prove to be correct we are not half way through as many commentators believe we are. It is not only on mortgages that we will see large losses going forward. It is also on consumer loans and credit cards reflecting the high and rising unemployment in the US. The corporate sector, outside banking, construction and real estate, which was well capitalised prior to the crisis, will also start to report large losses including bankruptcies as the crisis drags on. These estimates suggest that banks may still be in need of additional equity capital over and above what the recent US Treasury stress tests have suggested, see estimates in IMF (2009b). However, as noted also by the Fund, many governments have effectively through different guarantee provisions allowed for risk sharing between the banking sector and the government. In some cases, this is likely to limit the additional capital required.[16] In addition, if markets maintain the current optimistic view that we are over the worst, this may also help to bolster confidence and reduce the need for equity capital.

Total estimated losses on loans and securities originated in the Eurozone and the UK amount to $1,193 billion. The magnitude of these numbers shows that the credit crisis is also a European phenomenon. Numbers for Japan show that expected losses

---

15   This corresponds to 19 percent of US GDP in 2008.
16   IMF's estimates have not prevented equity markets from rallying. Thus, markets including financial stocks have soared from the beginning of March 2009 to the time of writing end May 2009. The above loss estimates raise the question whether the rally will be reversed should some of the projected losses turn out to be realised? During the crisis in the early 1930s, there were 6 bear market rallies, with equity prices increasing by more than 20 percent, which were all reversed, see Christiansen (2009).

on Japanese assets are small relative to the other regions.

How bad it will get depends crucially on the macro economy, which, however, is extremely difficult to forecast because we are in completely new territory with financial markets under stress and global stimulus packages including bailouts, which are hard to quantify in terms of their impact on the real economy. The IMF's outlook is based on the view that it takes long time to overcome a deep financial crisis.[17] The IMF is therefore expecting only a weak US recovery in 2010 with essentially a flat GDP profile relative to 2009. That said, other forecasters including JPMorgan have a more optimistic view on the speed of recovery. For the current year, JPMorgan and the IMF are roughly in line, but JPMorgan is expecting US GDP to grow by 2.7 percent in 2010. If the JPMorgan scenario materialises, losses are likely to be significantly smaller than losses estimated by the IMF.[18]

GDP forecasts				
	IMF		JPMorgan	
	2009	2010	2009	2010
USA	−2.8%	0.0%	−2.5%	2.7%
Eurozone	−4.2%	−0.4%	−4.1%	1.2%
Japan	−6.2%	0.5%	−6.8%	2.3%
China	6.5%	7.5%	7.2%	8.5%

Table 2. GDP forecasts. Source: IMF (2009b) and JPMorgan (2009).

High leverage amplifies the downward spiralling of markets and economies – policies can only mitigate this process

Large losses in the banking system have led banks to hoard cash in order to rebuild their balance sheets. This has led to an outright contraction of credits extended to the private sector, in spite of the infusion of public money, which essentially has sent all advanced economies into deep recessions. Thus, firms facing tougher financing conditions in advanced economies scaled back on capital spending and also started to lay off workers at a large scale in autumn 2008, when the credit crisis accelerated. Indebted consumers are also hit hard while they are, at the same time, suffering from sharp declines of their property and stock market wealth. In the US this has led consumers to hold back on spending and raise their savings. The household savings rate is now approaching 5 percent of disposable income, see figure 1. Relative to what is the average long-term prudent savings rate, consumers are about half-way through, but there is a long way to go in terms of rebuilding household wealth. Hedge funds and other highly geared investors have responded to the credit crunch by undertaking massive equity sell-offs, which accelerated the downward spiralling of prices of

---

17   This is the lesson of previous deep financial crises, see IMF (2009c).
18   In a discussion of this issue, it should also be noted that losses have so far been much higher than forecasted by the IMF.

risky assets in the fall of 2008.[19] The IMF (2009b) is projecting that private credit in advanced economies will decline not only in 2009 but also in 2010. Thus, the process of deleveraging banks' balance sheets and removing toxic assets is likely to be protracted given the very high leverage to begin with.[20] Regardless of massive asset swaps, including outright socialisation of financial institutions, extensive quantitative easing and record low policy interest rates, there is a limit to how much policy-makers can do to offset the deleveraging in the banking sector.[21]

Credit is therefore likely to be scarce also going forward. How scarce depends to a large extent on the macro economy. Under the optimistic JPMorgan scenario, banks will have significantly smaller losses on the corporate sector and on consumers and they will also be able to report much higher earnings. All of this will help mitigate the credit squeeze.

Global imbalances and the lack of international coordination and cooperation

Global imbalances and the lack of satisfactory international cooperation is also a feature of this crisis. Prior to the outbreak of the crisis in August 2007, leading international economists including international organisations warned that the magnitude of the US trade deficit was unsustainable. The US was advised to mitigate the problem by reducing consumption growth. China should appreciate its currency to help expand US exports to China and to reduce US imports from China. This policy advice was, however, ignored by the involved parties. Had the policy been adhered to, the adjustment of the world economy would have been less painful. Thus, the US consumer would have been less indebted, the savings ratio would have been higher and he would not need to cut his spending by as much as he is doing right now. China on the other hand would have had to rely more on domestic spending than exports in its development strategy, which would also have created a better balance relative to other regions, including Europe in particular. China would probably have had lower growth during the boom, but would also have avoided the severe consequence of a complete collapse of its exports as we have seen recently. This grand coordination plan did not have much effect on policies in spite of many high-level meetings between the US Treasury Secretary and officials from China. As noted by Dermine in this book, it is not only in macroeconomic coordination the world could do better. The supervision of international financial firms is lagging behind and much progress is needed in this area, though this was not the key cause of the crisis. A key problem in the field

---

19 The more leveraged the players are the worse is this vicious circle that has been studied extensively by Minsky (1977).

20 Extensive shadow banking, for example through special investment vehicles, is an important explanation of the excessive risk taking. See Dermine's chapter in this book.

21 The US Fed funds rate is in the interval between 0 and a quarter of a percent. The leading Eurozone policy interest rate is at 1 percent.

of international coordination and regulation is the unwillingness of sovereign states to delegate power to international institutions, that is, countries continue to play the Nash strategy instead of seeking truly international solutions, which is of particular importance in a time of globalised financial markets. The rise of China to become a superpower and the decline and influence of the United States does not help in making fast progress since progress in international politics normally requires strong leadership and coherence, including a common set of values.

Concluding comments

Crises are inevitable because long expansion periods almost always create excessive optimism and risk taking, that is, animal spirit to use a phrase of Keynes (1936). This leads to too much debt creation and also often to asset price inflation. Booming economies with overvalued asset prices and high debt are vulnerable to adverse shocks, including a more negative outlook, since this can easily trigger a sharp decline in asset prices. Plummeting asset prices lead banks to hold back on lending to reduce risks and to rebuild balance sheets, which pushes the economy further down the hole.

This boom-bust asset-price driven cycle is not new. What is unique this time is the magnitude of the leverage and risk taking in the banking sector in the US and Europe. This was facilitated by an extended period of low interest rates, extensive shadow banking and the development of new financial products, which turned out to be much more risky than first perceived. With the benefit of hindsight it is also clear that inadequate regulation is part of the picture. This is particularly obvious in the case of the US sub-prime market, which should not have been allowed to become so big, see also Shiller (2008). However, inadequate regulation and enforcement is also part of the problem in a number of other markets; see Dermine's analysis in this book and Soros (2008).

In spite of a *déjà vu* of past boom-bust asset price driven cycles, it is important to mention that there has been some progress in policy-making. Most importantly we avoided Great Depression II thanks to timely and forceful policy interventions led by the US Federal Reserve and the US Treasury; see also Eichengreen and O'Rourke (2009). The cost of this insurance against a meltdown of the financial sector and the world economy will be paid in terms of accelerating public deficits in all advanced economies. Moreover, the IMF projects a contraction in private credits in advanced economies also in 2010 due to a continued deleveraging of the banking sector, which the infusion of public money only mitigates but cannot reverse. The outlook is therefore for low growth and large losses in the banking sector. The exact magnitude of the losses depends crucially on the timing and recovery of advanced economies, which is unusually hard to forecast because we are in completely new territory.

From a macroeconomic policy point of view this crisis is likely to lead central banks to rethink the way they conduct monetary policy, see also IMF (2009a). The sole focus on inflation targeting has been shown to be inadequate, that is, there is a call for putting more weight on preventing asset market bubbles, though this is not an easy task since bubbles are hard to identify in the first place, and since political pressure could jeopardise attempts to prick bubbles. Progress in policy-making and the hard lessons learned from this crisis are likely to help in preventing large boom-bust cycles over the coming decade but it would be naive to think that it is a smooth ride from now on. Hypes in markets combined with a tendency to forget past lessons and a tendency to think that this time is different can always put the economy on an unsustainable path, which is eventually followed by a sharp downturn.

References

Akerlof, G.A. and R.J. Shiller, 2009, Animal Spirits: How Human Psychology Drives the Economy, and Why it Matters for Global Capitalism, Princeton University Press.

Buiter, W.H., 2008, Lessons from the North Atlantic financial crisis, Unpublished manuscript.

Christiansen, J., 2009, Aktiestigninger er et bear market rally [Stock market rebound is a bear market rally], LD Invest A/S.

Dermine, J., 2009, Avoiding international financial crises, Steen Thomsen, Caspar Rose and Ole Risager (eds.): Understanding the financial crisis: investment, risk and governance. SimCorp StrategyLab, 2009.

Eichengreen, B. and K.H. O'Rourke, 2009, A Tale of Two Depressions, Unpublished Manuscript.

Financial Times, Various issues, 2008–09.

International Monetary Fund, 2009a, Initial Lessons of the Crisis, February 2009, Washington DC.

International Monetary Fund, 2009b, Global Financial Stability Report, April 2009, Washington DC.

International Monetary Fund, 2009c, World Economic Outlook, April 2009, Washington DC.

JPMorgan, 2009, Global Data Watch, 22 May 2009.

Keynes, J.M., 1936, The General Theory of Employment, Interest, and Money, MacMillan, London.

Minsky, H., 1977, The Financial Instability Hypothesis, Nebraska Journal of Economics and Business.

Risager, O., 2009, Managing Macro Financial Risks, Journal of Applied IT and Investment Management, April 2009.

Roubini, N., 2008, The Coming Financial Pandemic, Foreign Policy March/April, 2008.

Shiller, R.J., 2008, The Subprime Solution: How Today's Global Financial Crisis Happened, and What to Do about It, Princeton University Press.

Soros, G., 2008, The New Paradigm for Financial Markets, Public Affairs.

Regulation and risk management

OVERVIEW

2 Regulation and risk governance

by Caspar Rose

 Caspar Rose, Ph.D., is a professor at Copenhagen Business School, Department of International Economics and Management and is also affiliated with the Center for Corporate Governance. He holds a master of laws as well as a doctorate in finance and his research focuses on operational risk management as well as corporate governance on which he has published extensively in international journals. Caspar Rose has worked in Dansk Industri (Confederation of Danish Industries) as special legal advisor as well as chief analyst at the Danske Bank, Group Operational Risk. He also serves as external consultant for financial institutions. Contact: car.int@cbs.dk

The financial crisis has revealed that the current legislative framework has proved insufficient. As a consequence, many commentators have argued that there is a need for new regulation in order to reduce the pro-cyclicality of the existing Basel Capital Accord solvency rules as well as avoiding moral hazard problems from deposit insurance. The use of off-balance items such as structured investment vehicles (SIV) as well as the lack of transparency regarding over-the-counter (OTC) transactions have also come in for particular scrutiny. The role of credit rating agencies facilitating complicated structured financial products without a proper due diligence has been heavily debated too. In other words, many of the fundamental building blocks in the global financial system have been seriously questioned following the crisis. However, perhaps most fundamentally, the existing regulatory framework as well as complicated risk management models failed to prevent the crisis that emerged from financial institutions' liquidity problems.

New regulation often follows a crisis

Financial institutions as well as financial markets have been heavily regulated for several years. It was started by the US with the introduction of the Securities Act of 1933 and the Securities and Exchange Act of 1934 following the crash of 1929. These acts, once enacted, were followed by depressed markets for several years. Since many investors had lost money, both houses of Congress conducted lengthy hearings to find the cause and the culprits. The hearings were marked by sensationalism and wide publicity and the two acts were the direct result of the congressional hearings. The two acts were later supplemented by several other laws, which were enacted following other financial crises; Weston, Mitchell and Mulherin (2004) provide an overview. The latest example is the comprehensive Sarbanes-Oxley Act of 2002, which deals with issues such as: enhanced financial disclosure, auditor independence, conflicts of interest, and corporate fraud among other things. Europe has followed the legislative initiatives from the US by implementing regulation in the form of various directives.

Today we are witnessing a similar situation with the current financial crisis where politicians feel strong public pressure to do something. In other words everyone expects that politicians as well as financial authorities will introduce new legislative initiatives combined with tougher enforcement of existing regulation. This prompts a crucial question: how should we go ahead and redesign the regulatory framework? As Dr. Hugo Banziger puts it in the final chapter of this book: "We cannot go back to the good old simple days." Dr. Banziger emphasises that future regulation should not be based on fear and control. Instead there is a need to create the right incentives through smart laws.

Key points by the contributors

John Hull proposes steps that can be taken by financial institutions and their regulators to improve the way risks are understood and managed in the future. Specifically, he argues that transparency is an important feature of a financial product. If the cash flows from a product cannot be calculated in a relatively straightforward way, the product should not be traded. Collateralisation should become a requirement in OTC markets. Compensation schemes should be chosen to better align the interests of traders with the long-term interests of the financial institutions they work for.

Jean Dermine argues that the set of reforms proposed recently by national and international groups to reduce the risk of global banking crises constitute an incomplete reform agenda. He says that there are three important issues to be addressed:
- accountability of bank supervisors and regulators;
- an end to the too-big-to fail doctrine;
- a satisfactory regime for the supervision and bailing out of international financial groups.

David Lando discusses the use of mathematical models in risk management. He argues that models can create scenarios and we can use them to better understand the consequences of these scenarios. Perhaps not down to the last penny, but they can be useful when they are within the right order of magnitude. The scenarios we choose to include are grounded in history and experience but they are not the outcome of a model. Management must decide, whether a scenario that involves a repeat of the Great Depression should be a real possibility in the model that is an event with non-vanishing probability. Management decides whether capital should be set aside to withstand such an event. If that choice is made, shareholders must understand that this causes the stock to be less risky and thus expected returns will need to be sacrificed.

Marc Schröter analyses one of the most debated subjects within enterprise architectures and IT solutions, the question of best-of-breed versus integrated solutions. In short, the advantage of best-of-breed solutions is that it can provide richer functionality to specific areas of the operating model. On the other hand, it introduces challenges in terms of multiple systems, duplicate entry of the same data, interfaces to other systems via middleware, redundant data storage and in general a more complex infrastructure.

The analysis reveals that no matter which operating model investment managers have, they should seek to have a core data master as part of their enterprise architecture to meet their strategic challenges. The core ensures that all information can be consolidated, is available online and that underlying data and calculations are consistent. The core should not only be a data repository, but should also contain business logic for calculating derived information and it should also contain functionality such as permissioning and audit trail to make it an operational system.

Why we cannot live without financial regulation

Ideally, the purpose of financial regulation is to internalise the social costs of potential bank failures via capital adequacy rules, the so-called Basel 'pillars', see Brunnermeier et al. (2009), or more precisely:

1 to constrain the use of monopoly power and distortions to competition;
2 to protect the essential needs of ordinary people in cases where information is hard or costly to obtain;
3 to reduce substantial externalities, that is that the total costs of market failure exceed both the private costs of failure and the extra costs of regulation.

Contrary to other markets for goods and services, financial markets cannot entirely rely on there being completely free markets because they are not sufficient to sustain the financial system. Financial markets participants deal with other people's money. Economists designate it the principal agency relationship, which may create moral hazard problems, as financial markets are characterised by substantial informational asymmetries.

Contrary to non-financial firms there are severe externalities if financial institutions fail. To illustrate, the failure of a bank weakens the other banks and financial markets with which they are involved, whereas failure of, say, a car manufacturer tends to strengthen the remaining car companies as a competitor is removed. If, however, Lehman Brothers fails, due to a maturity mismatch between liabilities and assets, then investors might believe that Merrill Lynch is similar to Lehman, so lenders and depositors will lose confidence and withdraw their funds. This creates a liquidity problem, which restricts Merrill Lynch's access to funds. Thus, interest rates increase, which spread to other banks and eventually to the real economy, see Brunnermeier (2009).

Bank failures may also have serious impacts on customers as nobody today can live without a bank. Externalities may consist of loss of access to future funding for the failed banks' customers. A client of a failed bank can always transfer their custom to surviving banks, but the new bank will have less direct information on this client so the start-up terms will be much tougher. Consequently, it is not a question of whether or not to have financial regulation, but instead how to regulate in a way that matters.

The current regulatory framework for risk management

The Basel Committee on Banking Supervision is a standing committee of the Bank of International Settlements (BIS). It formulates broad supervisory standards and guidelines and recommends statements of best practice in banking supervision such as the Basel II Accord, expecting that member authorities and other nations' authorities will take steps to implement them through their own national systems, including the EU, whether in statutory form or otherwise. The purpose of the committee is also to encourage convergence toward common approaches and standards. For more see www.bis.org/bcbs.

The EU Capital Requirements Directive (CRD) was formally adopted on 14 June 2006 building on the new Basel Accord. It affects banks and building societies and certain types of investment firms. The new framework consists of three pillars. Pillar 1 of the new standards sets out the minimum capital requirements firms will be required to meet for credit, market, and operational risk.

Moreover, the Basel Committee provides a forum for regular cooperation on banking supervisory matters. Its objective is to enhance understanding of key supervisory issues and improve the quality of banking supervision worldwide. It seeks to exchange information on national supervisory issues, approaches and techniques, with a view to promoting common understanding. At times, the Committee uses this common understanding to develop guidelines and supervisory standards in areas where they are considered desirable, see www.bis.org/bcbs/. In this regard, the Committee is best known for its international standards on capital adequacy; the Core Principles for Effective Banking Supervision; and the Concordat on cross-border banking supervision.

The Committee encourages contacts and cooperation among its members and other banking supervisory authorities. It circulates to supervisors throughout the world both published and unpublished papers providing guidance on banking supervisory matters. Contacts have been further strengthened by an International Conference of Banking Supervisors (ICBS) which takes place every two years.

On 29 May 2009, a proposal from the European Commission advocated the creation of a European Systemic Risk Council, to assess and warn about threats to financial stability in the region and a 'European System of Financial Supervisors' to oversee individual banks and financial firms. The European Systemic Risk Council would be headed by the president of the European Central Bank (ECB) and would include governors of the EU's 27 central banks. The panel would monitor broad risks in the region's financial system and intervene if necessary. Commission President Jose Manuel Barroso said, "It's now or never, if we cannot reform the financial sector when we have a real crisis, when will we?" – see SPEECH/09/273.

Operational risk becomes more visible during a crisis

While the management of operational risk has always been a fundamental element of banks' risk management procedures, Basel II introduced a new dimension in the form of separate capital requirements and heightened expectations for the management of operational risk. Improvements in the internal governance and other aspects of a bank's risk management and measurement framework are expected to coincide with the increased focus on operational risk. Operational risk is defined as 'the risk of loss resulting from inadequate or failed internal processes, people and systems or from external events,' see The Basel Committee on Banking Supervision, http://www.bis.org/bcbs/.

This definition includes legal risk, but excludes strategic and reputational risk. Legal risk is included, but is not limited to, exposure to fines, penalties, or punitive damages resulting from supervisory actions, as well as private settlements. The generic causes cover people, processes, systems and external events.

In times of economic recession or financial crises, there is a higher risk that certain operational risk events materialise such as rogue trading, fraud and misselling/ incorrect advice. For instance, asset managers may be tempted to manipulate their earnings and financial statements to avoid reporting huge losses, see Rose (2009).

To illustrate, Société Générale, one of the largest banks in Europe, was thrown into turmoil in January 2008 after the chairman revealed that a rogue employee had executed a series of 'elaborate, fictitious transactions' that cost the company more than $7 billion, the biggest loss ever recorded in the financial industry by a single trader. Furthermore, on 10 December 2008, FBI agents arrested the former chairman of NASDAQ Bernard Madoff and charged him with one count of securities fraud.

According to federal charges, Madoff said that his firm has "liabilities of approximately $50 billion." Banks from outside the U.S. have announced that they have potentially lost billions in dollars as a result, see http://en.wikipedia.org/wiki/Bernard_Madoff#cite_note-SEC1-11.

Moreover, customers who have seen their returns diminish substantially due to an economic recession may sue financial institutions claiming that they have received incorrect advice from asset managers, including failure to assess product suitability. In times of an economic recession where the likelihood of default increases, recovery of collateral becomes pivotal. Therefore, it is important that the internal legal processes that are in place to ensure ownership of collateral are robust and checked carefully. A small legal technicality may be crucial as recovery may turn out to be legally impossible.

The many faces of the regulatory failure

The regulatory failure has manifested itself in several ways. However, there are a number of large failures that have especially attracted their attention. First, the legal vacuum of credit default swaps (CDS) has been criticised by many commentators. A CDS involves a buyer who makes periodic payments to a seller, and in return receives a payout if an underlying financial instrument defaults. CDS contracts have been compared with insurance contracts because the buyer pays a premium and in return, receives a sum of money if one of the specified events occurs. However, there are a number of differences between CDS and insurance. For example a seller need not be a regulated entity and a buyer of a CDS does not need to own the underlying security or other form of credit exposure. In fact the buyer does not even have to suffer a loss from the default event, see www.wikipedia.org/wiki/Credit_default_swap for a description.

CDSs allow investors to speculate on changes in an entity's credit quality, since generally CDS spreads will increase as credit-worthiness declines and decline as credit-worthiness increases. Despite the fact that the CDS market was enormous it remained completely unregulated. The market activity in 2008 amounted $55 trillion and it had significant effects on ordinary regulated markets, see www.wikipedia.org/wiki/Credit_default_swap. The CDS market was characterised by herd behaviour and banks' top management did not always fully understand the product's risks. Or as Tom Wilson, Chief Risk Officer at Allianz SE says, "when the music started to play people started to dance", see the final chapter of this book.

Moreover, the lack of transparency in these instruments contributed significantly to uncertainty in the market. A main reason is that there is no legal requirement for disclosure surrounding CDSs, either to the US capital markets regulator, the Securities and Exchange Committee (SEC), or any other agency. As a consequence, SEC in 2008 worked with the financial services industry to develop one or more central counterparties, clearance, and settlement systems, and trading platforms for this market. SEC will now work with Congress to make CDSs more transparent while "giving regulators the power to rein in fraudulent or manipulative trading practices" and help ordinary investors to better assess the risks involved. The EU is currently contemplating how to strengthen the supervision of financial institutions.

There is no doubt that the regulation of credit rating agencies failed. This is a very serious problem, since rating agencies had played a key role in recent years especially with mortgage backed securities (MBS) and collateralised debt obligations (CDO), see www.wikipedia.org/wiki/Credit_rating_agency. After assigning high ratings to these products, including sub-prime, ones between 2004–2007 rating agencies began to a rapidly downgrade these instruments. This not surprisingly raised the question about the quality of the credit ratings with regard to structured products. It is a fact that high ratings encouraged investors to buy securities backed by sub-prime mortgages, helping finance the housing boom. The reliance on agency ratings and the way ratings were used to justify investment decisions led many investors to treat securitised products, some based on sub-prime mortgages, as equivalent to higher quality securities, see p. 32 in Financial Stability Forum (2008).

This was exacerbated by the SEC's removal of regulatory barriers and its reduction of disclosure requirements. Critics allege that the rating agencies suffered from conflicts of interest, as they were paid by investment banks and other firms that organise and sell structured securities to investors. The SEC has now approved measures to strengthen oversight of credit rating agencies, following a ten-month investigation that found "significant weaknesses in ratings practices," including conflicts of interest. See also SEC (2008) 'Summary Report of Issues Identified in the Commission's Staff's Examinations of Select Credit Rating Agencies'.

This naturally raises the question of whom to blame? Between Q3 2007 and Q2 2008, rating agencies lowered the credit ratings of $1.9 trillion in MBSs. Financial institutions felt they had to lower the value of their MBS and acquire additional capital so as to maintain capital ratios. If this involved the sale of new shares of stock, the value of the existing shares was reduced. Thus ratings downgrades lowered the stock prices of many financial firms. Arnold Kling said to the Congress that a high-risk loan could be "laundered" by Wall Street and returned to the banking system as a highly rated security for sale to investors, obscuring its true risks and avoiding capital reserve requirements *(Financial Times)*.

As a result, following the findings of the Financial Stability Forum one may argue that the credit rating agencies should:
- provide disclosure on their due diligence performed on the underlying assets;
- disclose their track records as a test of their rating capabilities;
- offer mandatory information requirements emphasising that investors should not over-rely on ratings.

Naked short selling has become a major problem for regulators especially in the US where it poses a serious threat for the trust of financial markets. Naked short selling entails that sellers do not borrow stocks before an execution of a short sale, a practice that becomes evident once the stock is not delivered. Such trades can generate unlimited orders, overwhelming buyers and drive down prices. Failure to deliver can be used to manipulate markets. Harvey Pitt, a former SEC chairman calls it "fraud". There are strong indications that naked short selling contributed to the fall of both Lehman Brothers and Bear Stearns *(Financial Times)*.

Until recently the SEC has been quite reluctant to act. An SEC report showed that of about 5,000 e-mailed tips related to naked short selling received from January 2007 to June 2008, 123 were forwarded for further investigation, but none led to actions (Bloomberg). Kotz, who is inspector general in the SEC, said that the enforcement division "is reluctant to expand additional resources to investigate complaints".

The following situations illustrate the magnitude of the problem with naked short selling. On 30 June 2008, someone started a rumor that Barclays Plc. was ready to buy Lehman for 25 percent less than the day's price. The purchase did not materialise. On the previous trading day, 27 June, the number of shares sold without delivery jumped to 705,103 from 20,690, a 23-fold increase on 26 June. On 17 September, two days after Lehman Brothers filed for Chapter 11 bankruptcy protection, the number of failed trades climbed to 49.7m, 23 percent of overall volume in the stock according to Bloomberg. The next day SEC announced a ban on shorting financial companies, which several jurisdictions around the world decided to follow, see SEC Press Release 211 (2008).

Following such instances, it is not surprising that the SEC at the start of 2009 is playing tough. For instance, the SEC Enforcement Division has announced what will be the largest settlements in the history of the SEC for investors who bought auction rate securities from Citigroup, UBS, Merrill Lynch and Bank of America. The Commission has finalised the Citi and UBS settlements. It has charged a number of Wall Street brokers with defrauding their customers when making more than $1 billion in unauthorised purchases of sub-prime-related auction rate securities. The SEC has also charged five California brokers for pushing homeowners into risky and unsustainable sub-prime mortgages, and then fraudulently selling them securities that were paid for with the mortgage proceeds. More importantly, it has charged Fannie Mae and Freddie Mac with accounting fraud in 2006 and 2007 respectively, and the companies paid more than $450m in penalties to settle the SEC's charges, see SEC Press Release 211 (2008).

The ban on short selling may be seen as a bit of a paradox as modern textbook finance theory shows that short sale is not merely relevant if seller expects the shares to decline in value and wishes to profit from the decline. Another reason is to decrease the sensitivity of a portfolio to market movements, see Elton and Gruber (1995). Since the return on short sales is the opposite of the return on a long position, a portfolio that includes short sales as well as long positions reduces the exposure to market movements.

The ability to short sell is a key ingredient in many theoretical asset pricing models, see Huang and Litzenberger (1988), as well as more simple models such as capital asset pricing models (CAPM) that are used extensively in practice. This is indeed a paradox as prohibiting short sales may question if we can still rely on our fundamental financial asset pricing models such as CAPM where short is needed in order to create a more risky portfolio with a higher expected return. Moreover, introducing short sales constraints create additional challenges for the theoretical asset pricing models, see Duffie (1996) for an overview.

How to deal effectively with systemic risk?

The current approach to systemic regulation implicitly assumes that we can make the system as a whole safe by simply trying to make sure that individual banks are safe. However, in trying to make themselves safer, banks and other highly leveraged financial institutions can behave in a way that collectively undermines the system, see Brunnermeier et al. (2009).

The problem of the present Basel rules is that liquidity problems may create serious solvency problems. To avoid liquidity problems the bank may often be forced to sell assets. Such sales will drive down the current market price of the same assets held by other banks when valued on a mark-to-market basis. Attempts to deal with a liquidity problem leads to a decline in asset values thereby creating a solvency problem, hence solvency is not exogenous to liquidity. A liquidity spiral entails that falling asset markets lead banks, investment houses and others to make more sales, to 'deleverage', which further drives down asset prices resulting in amplifying a dynamic process with reductions in credit expansion and the interbank market dries out.

Therefore, there is need for countervailing force to the natural decline in measured risks in a boom and the subsequent collapse which should be as rigorous as possible. Regulators should increase the existing solvency requirements for example in the case of above average growth of credit expansion and leverage. This might not eliminate the cycle, but it might help to stabilise asset prices, see Brunnermeier et al. (2008) for an analysis.

However, if a financial crisis is primarily caused by endogenous risks, we may question our existing models. One implication is that the standard format of banking stress tests is fundamentally insufficient which has also been noticed by Hugo Banziger, CRO at Deutsche Bank. These stress tests which are mandatory under Basel II, review the effect on each bank's profits and capital of historically based, exogenous shocks. However, such tests, focusing on exogenous risk, will miss out on the more important second and higher round effects.

Stress testing is a critical tool used by banks as part of their internal risk management and capital planning. It is also a key component of the supervisory assessment process to identify vulnerabilities and assess the capital adequacy of banks. The principles establish expectations for the role and responsibilities of supervisors when evaluating banks' stress testing practices.

The Basel Committee on Banking Supervision on 20 May 2009 issued its 'Principles for sound stress testing practices and supervision.' The paper sets out a comprehensive set of principles for the sound governance, design and implementation of stress testing programmes at banks. The principles seek to address the weaknesses in banks' stress tests that were highlighted by the financial crisis, see the Basel announcement on www.bis.org/press/p090520.htm.

"The current crisis underscores the importance of stress testing as an essential risk management and capital planning tool", noted Mr Nout Wellink, Chairman of the Basel Committee on Banking Supervision and President of the Netherlands Bank. He added, "stress testing programmes should be fully integrated in banks' governance frameworks, and the Basel Committee will follow up to ensure that these principles are implemented." See www.bis.org/press/p090520.htm.

Mr Klaas Knot, Chairman of the Basel Committee's Risk Management and Modelling Group and Division Director of Supervision Policy at the Netherlands Bank, said: "Stress testing plays an important role in strengthening not only bank corporate governance but also the resilience of individual banks and the financial system." He explained that "the Basel Committee's principles were developed with the long-term objective of deepening and strengthening banks' stress testing practices and supervisory assessment of those practices." Specifically, the Basel Committee recommends that "Stress testing programmes should cover a range of scenarios, including forward-looking scenarios, and aim to take into account system-wide interactions and feedback effect."

The Basel Committee states that, "When analysing the potential impact of a set of macroeconomic and financial shocks, a bank should aim to take into account system-wide interactions and feedback effects. Recent events have demonstrated that these effects have the capacity to transform isolated stress events into global crisis threatening even large, well-capitalised banks, as well as systemic stability. As they occur rarely, they are generally not contained in historical data series used for daily risk management. A stress test supplemented with expert judgement can help to address these deficiencies in an iterative process and thereby improve risk identification."

Unfortunately, any simple attempts to adjust stress test for endogenous risk have not yet borne fruit, but there is reason to believe that this interesting issue will be pursued among academics in the coming years.

The same line of argument can be made regarding existing market risk models. As noted by Hull, "There is a tendency for Value at Risk calculations to assume that correlations observed in normal market conditions will apply in stressed marked conditions."

Critics of the Basel pillars argue that the requirements for capital adequacy are pro-cyclical. Therefore, there is a need for counter-cyclical measures that are tough during a credit boom and more relaxed during a crisis. As Lars Rohde, CEO of ATP (see the final chapter of this book) notes, there is a need for board members to have a 'mental emergency plan' to handle large systemic risks. The rules appeared to do too little to limit bankers credit expansion in the boom, nor to help offset the wave of panic failures and deleveraging in the subsequent crisis. The requirements do not contain a detailed analysis of incentives and sanctions. They only suggest preferred forms of bank behaviour and do not effectively address the too big to fail problem. This has been recognised by Bernanke among others who seem to be hinting that the Basel rules on capital adequacy should be re-examined. "We should revisit capital regulations, accounting rules, and other aspects of the regulatory regime to ensure that they do not induce excessive pro-cyclicality in the financial system and the economy." He tells his audience that it's unacceptable that some firms are "too big to fail" (*Telegraph*, 13 January 2009).

How to move forward

The current regulatory framework has been heavily criticised following the crisis especially due to a number of spectacular financial scandals. As US Treasury Secretary Hank Paulson says, "The cost to our nation will be even larger if we do not overhaul our regulatory system," see *Financial Times*. AIG was the largest insurance firm in the world when it reported a loss of nearly $170 billion in March 2008 which affected several other large financial institutions. Bernanke told the Senate that AIG *de facto* served as a hedge fund. It had made a huge amount of irresponsible bets and losses due to a failure in financial regulation. President Barack Obama's administration reacted angrily to news that the AIG group, which is now 80%-owned by US taxpayers after it received federal bailouts of more than $150 billion, is to pay $165m in bonuses to top executives. The bulk of the bonuses are expected to go to staff in AIG's London-run financial products unit, the very business division that got the institution into trouble last year. Lawrence Summers, Obama's senior economic adviser, called AIG's behaviour "outrageous", but said that the company was obliged to pay because of legally binding contracts made during the boom years, see *Financial Times*.

This raises the crucial questions: should the government always pick up the pieces even if it creates a moral hazard problem and how should we design a new type of regulation that avoids similar events without creating excessive overregulation?

Ideally, I find that the goals of new regulation should concentrate on:
- fostering market stability while maintaining capitalism;
- creating regulation that facilitates on long-term incentives;
- building a regulatory structure organised by objectives;
- focusing on risk across the financial system, minimising systemic risk;
- fostering increased transparency, which is necessary to fully understand financial products and their associated risks;
- avoiding excessive rules and bureaucracy as the lessons from the Sarbanes-Oxley Act have shown that massive new rules may not offer any real degree of protection;
- producing rules based on substance rather than form;
- providing regulators with the right incentive structures as well as independence.

References

Basel Committee on Banking Supervision, c.f. http://www.bis.org/bcbs/.

Basel Committee (2009) 'Sound stress testing principles issued by Basel Committee' 20 May 2009, see http://www.bis.org/press/p090520.htm.

Bloomberg

Brunnermeier, Markus, Andrew Crocket, Charles Goodhart, A.D. Persaud and Hyun Shin, 'The Fundamental Principles of Financial Regulation' (2009), Geneva Reports on the World Economy 11.

Duffie, Darrell, (1996), 'Dynamic Asset Pricing Theory', Princeton.

Elton, Edwin and Martin Gruber (1995) 'Modern Portfolio Theory and Investment Analysis', fifth ed. *John Wiley and Gruber.*

Financial Times

Huang Chi-fu and Robert H. Litzenberger (1988), 'Foundations for Financial Economics'.

J. Fred Weston, Mark L. Mitchell and J. Harold Mulerin (2004), 'Takeovers, Restructuring and Corporate Governance', forth ed. Printice Hall.

Financial Stability Forum, 'Report of the Financial Stability Forum on Enhancing Market and Institutional Resilience', 7 April 2008.

'Principles for sound stress testing practices and supervision', The Basel Committee on Banking Supervision, May 2009.

Rose, Caspar (2009) 'New Challenges for Operational Risk after the Financial Crisis', *Journal of Applied IT and Investment Management* Vol.1 pp 27–31.

SEC (2008) 'Summary Report of Issues Identified in the Commissions Staffs Examinations of Select Credit Rating Agencies'.

SEC Press Release 211 (2008) SEC Halts Short Selling of Financial Stocks to Protect Investors and Markets see http://www.sec.gov/news/press/2008/2008-211.htm.

SPEECH/09/273 by EU President Jose Manuel Barroso, Brussels, 27 May 2009.

Wikipedia: http://en.wikipedia.org/wiki/Bernard_Madoff#cite_note-SEC1-11

Wikipedia: http://en.wikipedia.org/wiki/Credit_default_swap

Wikipedia: http://en.wikipedia.org/wiki/Credit_rating_agency

About the author

John Hull is Maple Financial Professor of Derivatives and Risk Management at the Joseph L. Rotman School of Management, University of Toronto. He has written three books 'Risk Management and Financial Institutions' (now in its second edition), 'Options, Futures, and Other Derivatives' (now in its seventh edition) and 'Fundamentals of Futures and Options Markets' (now in its sixth edition). The books have been translated into many languages and are widely used in trading rooms throughout the world. He has won many teaching awards, including University of Toronto's prestigious Northrop Frye award, and was voted Financial Engineer of the Year in 1999 by the International Association of Financial Engineers. He is co-director of Rotman's Master of Finance program. Contact: hull@ rotman.utoronto.ca

3
The credit crisis of 2007 and its implications for risk management

ABSTRACT This chapter considers the lessons for risk managers from the credit crisis of 2007. It explains the events leading to the crisis and the products that were created from residential mortgages. It discusses the reasons why the risks in these products were underestimated. It proposes steps that can be taken by financial institutions and their regulators to improve the way risks are understood and managed in the future.

3 The credit crisis of 2007 and its implications for risk management[1]

by John Hull

The risk management profession has come under a great deal of criticism since July 2007. How could it not have anticipated the meltdown in financial markets? Are the tools that it has developed in the last 20 years worthless? Was it too focused on historical data and not taking enough account of changing market conditions?

Although there is much that risk managers can learn from the credit crisis, many of the criticisms are unfair. Some risk management groups did realise that excessive risks were being taken during the period leading up to July 2007. At Merrill Lynch for example, Keishi Hotsuki, co-head of risk management, and David Rosenberg, chief North American economist, are on record as urging their employer to reduce its exposure to sub-prime mortgages. The reality of being a risk manager is that in good times, or times that appear to be good, you are usually ignored as being irrelevant. But, when disaster strikes, you get blamed.

This chapter builds on the analysis in Hull (2008, 2009a, 2009b) and Hull and White (2009). It examines the origins of the sub-prime crisis and the nature of the credit derivatives that were created. It looks at how the risks in credit derivatives were assessed and suggests areas that risk managers, the financial institutions they work for, and the bodies that regulate those financial institutions should pay more attention to in the future.

The origins of the crisis

The first decade of the 21st century has been characterised by a period of leveraging that lasted from 2000 to 2006 and a period of deleveraging that lasted from 2007 to 2009. During the leveraging period individuals and businesses found it relatively easy to borrow funds to buy assets. As a result, asset prices increased. During the deleveraging period there was a scramble to sell assets and prices declined.

The US housing market, which is the root cause of the sub-prime crisis, exemplifies the leveraging/deleveraging periods well. Figure 1 shows the S&P/Case-Shiller composite-10 index for house prices in the US between January 1987 and February 2009. In about the year 2000, house prices started to rise much faster than they had in the previous decade. The very low level of interest rates between 2002 and 2005 was an important contributory factor, but the increase was for the most part fuelled by mortgage lending practices.

---

1   The author is grateful to Richard Cantor and Roger Stein for useful comments on an earlier draft. All views expressed are his own.

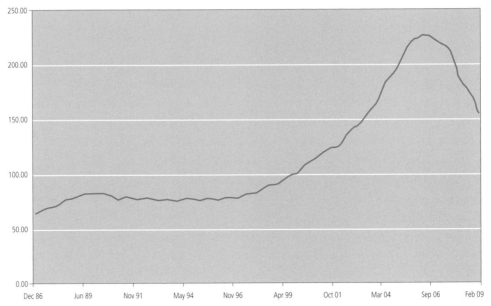

Fig. 1. S&P/Case-Shiller Composite-10 Index of US Residential Real Estate, 1987 to February 2009.

Mortgage lenders started to relax their lending standards in about 2000. This made house purchase possible for many families that had previously been considered to be not sufficiently creditworthy to qualify for a mortgage. These families increased the demand for real estate and prices rose. The mortgages were classified as sub-prime because they were much more risky than average. To mortgage brokers and mortgage lenders the combination of more lending and higher house prices was attractive. More lending meant bigger profits. Higher house prices meant that the mortgages were well covered by the underlying collateral. If the borrower defaulted, the resulting fore-closure would not lead to a loss.

To keep house prices rising, lenders had to find ways of continuing to attract new entrants to the housing market. They did this by continuing to relax their lending standards. The amount lent as a percentage of the house price increased. Adjustable rate mortgages (ARMs) were developed where there was a low 'teaser' rate of inter-est that would last for two or three years. Lenders also became more cavalier in the way they reviewed mortgage applications. Indeed, the applicant's income and other information reported on a mortgage application were frequently not checked. If house prices increased, lenders expected borrowers to refinance at the end of the teaser rate period. This would be profitable for the lenders because prepayment penalties were relatively high on sub-prime mortgages.

Mian and Sufi (2008) have carried out research confirming that there was a relax-ation of the criteria used for mortgage lending. Their research defines 'high denial zip codes' as zip codes where a high proportion of mortgage applicants had been turned down in 1996, and shows that mortgage origination grew particularly fast for these zip codes between 2000 to 2007. Moreover, their research shows that lending criteria were relaxed progressively through time rather than all at once because originations in high denial zip codes are an increasing function of time during the 2000 to 2007 period. Zimmerman (2007) provides some confirmation of this. He shows that subsequent default experience indicates that mortgages made in 2006 were of a lower quality than those made in 2005 and these were in turn of lower quality than the mortgages made in 2004.

Standard & Poor's has estimated that sub-prime mortgage origination in 2006 alone totalled $421 billion. AMP Capital Investors estimate that there was a total of $1.4 trillion of sub-prime mortgages outstanding in July 2007.

Residential mortgages in the US are non-recourse in many states. This means that, when there is a default, the lender is able to take possession of the house, but cannot seize other assets of the borrower. The result of this is that the borrower owns an American-style put option. He or she can at any time sell the house to the lender for the principal outstanding on the mortgage. (During the teaser-interest-rate period this principal often increased, making the option more valuable.) Not surprisingly, many speculators attracted by the availability of non-recourse financing for close to 100% of the cost of a house, were active in the residential real estate market. The purchase of a house was a 'no-lose' proposition. If house prices went up, a speculator could sell the house at a profit; if house prices declined, the speculator could exercise the put option.

As figure 1 illustrates, the house-price bubble burst during the 2006–2007 period. Many mortgage holders found they could no longer afford mortgages when teaser rates ended. Foreclosures increased. House prices declined. This led to other homeowners and speculators finding themselves in a negative equity position. They exercised their put options and 'walked away' from their houses and their mortgage obligations. This reinforced the downward trend in house prices. Belatedly mortgage lenders realised that the put options implicit in their contracts were liable to be very costly.

The United States was not alone in experiencing declining real estate prices. Prices declined in many other countries as well. The United Kingdom was particularly badly affected.

The products that were created

The originators of mortgages in many cases chose to securitise mortgages rather than fund the mortgages themselves. Securitisation has been an important and useful tool in financial markets for many years. It underlies the 'originate-to-distribute' model that was widely used by banks prior to 2007.

Securitisation played a part in the creation of the housing bubble. Research by Keys et al (2008) shows that there was a link between mortgage securitisation and the relaxation of lending standards. When considering new mortgage applications, the question was not, "Is this a credit we want to assume?" Instead it was, "Is this a mortgage we can make money on by selling it to someone else?"

When mortgages were securitised, the only useful information received about the mortgages by the buyers of the products that were created from them was the loan-to-value ratio (that is, the ratio of the size of the loan to the assessed value of the house) and the borrower's FICO score.[2] The reason why lenders did not check information on things such as the applicant's income, the number of years the applicant had lived at his or her current address, and so on was that this information was considered irrelevant. The most important thing for the lender was whether the mortgage could be sold to others, and this depended primarily on the loan-to-value ratio and the applicant's FICO score.

It is interesting to note in passing that both the loan-to-value ratio and the FICO score were of doubtful quality. The property assessors who determined the value of a house at the time of a mortgage application sometimes succumbed to pressure from the lenders to come up with high values. Potential borrowers were sometimes counselled to take certain actions that would improve their FICO scores.

We now consider the products that were created from the mortgages.

Asset backed securities

The main security created from pools of mortgages was an asset backed security (ABS). Figure 2 shows a simple example illustrating the features of an ABS. A portfolio of risky income-producing assets is sold by the originators of the assets to a special purpose vehicle (SPV) and the cash flows from the assets are allocated to tranches. In figure 2, there are three tranches: the senior tranche, the mezzanine tranche,

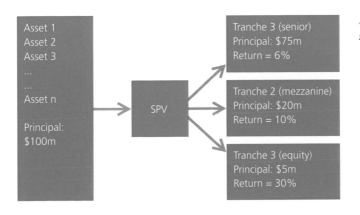

Fig. 2. An asset backed security (simplified)

---

2 The Fair Isaac Corporation, known as FICO, created the first credit scoring system in 1958, for American Investments, and the first credit scoring system for a bank credit card in 1970, for American Bank and Trust.

and the equity tranche. The portfolio has a principal of $100m. This is divided as follows: $75m to the senior tranche, $20m to the mezzanine tranche, and $5m to the equity tranche. The senior tranche is promised a return of 6%, the mezzanine tranche is promised a return of 10%, and the equity tranche is promised a return of 30%.

The equity tranche is much less likely to realise its promised return than the other two tranches. An ABS is defined by specifying what is known as a 'waterfall'. This defines the rules for allocating cash flows from the income-producing assets to the tranches. Typically, cash flows from the assets during a particular time period are allocated to the senior tranche until the senior tranche has received its promised return. Assuming that the promised return to the senior tranche is made in full, cash flows are then allocated to the mezzanine tranche. If the promised return to the mezzanine tranche is made in full and cash flows are left over, they are allocated to the equity tranche. The precise waterfall rules are outlined in a legal document that is usually several hundred pages long. If any cash flows remain after the equity tranche holders have received their promised returns they are usually used to repay principal on the senior tranche.

When an ABS is created from mortgages, it typically lasts for the whole life of the mortgages. The weighted average life of the mortgage pool depends on prepayments and defaults. At some stage, principal payments are made to tranches. The extent to which the tranches get their principal back depends on losses on the underlying assets. The first 5% of losses are borne by the principal of the equity tranche. If losses exceed 5%, the equity tranche loses its entire principal and some losses are borne by the principal of the mezzanine tranche. If losses exceed 25%, the mezzanine tranche loses its entire principal and some losses are borne by the principal of the senior tranche.

There are therefore two ways of looking at an ABS. One is with reference to the waterfall rules. Cash flows go first to the senior tranche, then to the mezzanine tranche, and then to the equity tranche. The other is in terms of losses. Losses of principal are first borne by the equity tranche, then by the mezzanine tranche, and then by the senior tranche.

The ABS is designed so that the senior tranche is rated AAA/Aaa. The mezzanine tranche is typically rated BBB/Baa. The equity tranche is typically unrated. Unlike the ratings assigned to bonds, the ratings assigned to the tranches of an ABS are what might be termed 'negotiated ratings'. The objective of the creator of the ABS is to make the senior tranche as big as possible without losing its AAA/Aaa credit rating. (This maximises the profitability of the structure.) The ABS creator examines information published by rating agencies on how tranches are rated and may present several structures to rating agencies for a preliminary evaluation before choosing the final one.

ABS collateralised debt obligations

Finding investors to buy the senior AAA-rated tranches created from sub-prime mortgages was not difficult. Equity tranches were typically retained by the originator of the mortgages or sold to a hedge fund. Finding investors for the mezzanine tranches was more difficult. This led financial engineers to be creative; arguably too creative. Financial engineers created an ABS from the mezzanine tranches of different ABSs that were created from sub-prime mortgages. This is known as an ABS collateralised debt obligation (CDO) or mezzanine ABS CDO and is illustrated in figure 3.

The senior tranche of the ABS CDO is rated AAA/Aaa. This means that the total of the AAA-rated instruments created in the example that is considered here is 90% (75% plus 75% of 20%) of the principal of the underlying mortgage portfolio. This seems high but, if the securitisation was carried further with an ABS being created from the mezzanine tranches of ABS CDOs, and this did happen, the percentage would be pushed even higher.

In the example in figure 3, the AAA-rated tranche of the ABS would probably be downgraded in the second half of 2007. However, it is likely to receive its promised return if losses on the underlying mortgage portfolio are less than 25% because all losses of principal would then be absorbed by the more junior tranches. The AAA-rated tranche of the ABS CDO in figure 3 is much more risky. It will get paid the promised return if losses on the underlying portfolio are 10% or less because in that case mezzanine tranches of ABSs have to absorb losses equal to 5% of the ABS principal or less. As they have a total principal of 20% of the ABS principal, their loss is at most 5/20 or 25%. At worst this wipes out the equity tranche and mezzanine tranche of the ABS CDO, but leaves the senior tranche unscathed.

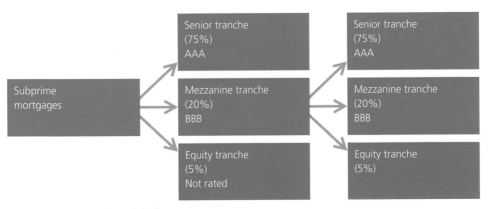

Fig. 3. An ABS CDO (simplified)

The senior tranche of the ABS CDO suffers losses if losses on the underlying portfolios are more than 10%. Consider, for example, the situation where losses are 20% on the underlying portfolios. In this case, losses on the mezzanine tranches are 15/20 or 75% of their principal. The first 25% is absorbed by the equity and mezzanine tranches of the ABS CDO. The senior tranche of the ABS CDO therefore loses 50/75 or 67% of its value. These and other results are summarised in table 1.

Losses to sub-prime portfolios	Losses to mezzanine tranche of ABS	Losses to equity tranche of ABS CDO	Losses to mezzanine tranche of ABS CDO	Losses to senior tranche of ABS CDO
10%	25%	100%	100%	0%
15%	50%	100%	100%	33.3%
20%	75%	100%	100%	66.7%
25%	100%	100%	100%	100%

Table 1. Losses to AAA tranches of ABS CDO in figure 3

ABSs and ABS CDOs in practice

The examples we have considered so far illustrate how the products that were created worked. In practice more than three tranches were created at each level of securitisation and many of the tranches were quite thin (in other words corresponded to a narrow range of losses). Figure 4 is an illustration, taken from Gorton (2008), of the type of structures that were created. The original source of the illustration is an article by UBS.

Two ABS CDOs are created in figure 4. One is created from the BBB rated tranches of ABSs (similarly to the ABS CDO in figure 3); the other is from the AAA, AA, and A tranches of ABSs. There is also a third level of securitisation based on the A and AA tranches of the mezzanine ABS CDO.

Many of the tranches in figure 4 (for example, the BBB tranches of the ABS that cover losses from 1% to 4% and are used to create the mezzanine ABS CDO) appear to be very risky. In fact, they are less risky than they appear when the details of the waterfalls of the underlying ABSs are taken into account. In the arrangement in figure 4, there is some over-collateralisation with the face value of the mortgages being greater than the face value of the instruments that are created by the ABSs. There is also what is termed 'excess spread'. This means that the weighted average of the returns promised to tranche holders is less than the weighted average of the interest due on the mortgages. If, because of these features, there are cash flows left over when all ABS tranches have received their promised returns, the cash flows are used to reduce the principal on the senior tranches.

Many banks have lost money investing in the senior tranches of ABS CDOs. The investments were typically financed at LIBOR and promised a return quite a bit higher than LIBOR. Because they were rated AAA, the capital requirements were minimal. In July 2008 Merrill Lynch agreed to sell senior tranches of ABS CDOs that had previously been rated AAA and had a principal of $30.6 billion to Lone Star Funds for 22 cents on the dollar.[3]

---

3   Merrill Lynch agreed to finance 75% of the purchase price. When the value of the tranches fell below 16.5 cents on the dollar, Merrill Lynch found itself owning the assets again.

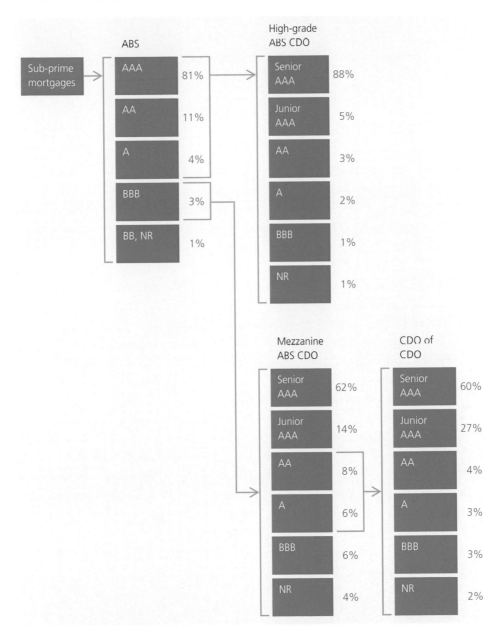

Fig. 4. A more realistic example of the structures created from mortgages

The rating agencies played a key role in the securitisation of mortgages. The traditional business of rating agencies is of course the rating of bonds. This is based largely on judgment. The rating process for instruments such as the tranches of ABSs and ABS CDOs was different from the rating of bonds because it was based primarily on models. The rating agencies published their models. Interestingly, different rating agencies used different criteria. Moody's criterion was expected loss. If a tranche's expected loss corresponded to the range of expected losses for a Aaa bond, it was rated Aaa. The S&P criterion was probability of loss. If the probability of loss for a tranche corresponded to the range of probabilities of default for a AAA bond, it was rated AAA.

Risk management

The promised return on the AAA/Aaa tranches of ABSs and ABS CDOs was higher than that on other assets rated AAA/Aaa. This should have served as a warning signal to financial institutions and their risk management groups. Financial markets do not usually increase expected return without there being an increase in risk. A reasonable assumption for risk managers is that there must be some risk that was not being taken into account by the models used by the rating agencies.

The role of models

Many financial institutions relied on the AAA/Aaa ratings of the senior tranches of ABSs and ABS CDOs without developing their own models or carefully examining the assumptions on which the models of the rating agencies were based. This was an unfortunate mistake. Only by developing a model, and critically examining the assumptions underlying the model, can a financial institution fully understand the nature of the risks being taken.

Without constructing a model, it is tempting to assume that a 25% to 100% tranche created from the BBB-rated tranches of an ABS has the same risk as a 25% to 100% tranche created from BBB bonds. However, as Hull and White (2009) show, even a simple model very quickly reveals that this is not the case. Although the BBB tranches do not have a higher expected loss or probability of loss than bonds that are rated BBB, the probability distribution of the loss is quite different, and this has implications for the riskiness of the tranches created from them.

Two key factors determine the riskiness of the tranches of a mezzanine ABS CDO. These are the thickness of the underlying BBB tranches and the extent to which defaults are correlated across the different pools used to create the mezzanine ABS CDO. Consider an extreme situation where the underlying BBB tranches are very thin and the mortgages in all pools are perfected correlated so that they have the same default rate. The BBB tranches of all the underlying ABSs then have a probability, say p, of being wiped out and a probability $1-p$ of being untouched. All tranches of the ABS CDO are then equally risky. They also have a probability p of being wiped out and a probability $1-p$ of being untouched. It would be appropriate to give all of them the same rating as the underling BBB tranches. This is an extreme example, but it illustrates why BBB tranches of ABS CDOs should not be treated in the same way as BBB bonds in the second level of securitisation.

When a model is constructed, it is tempting to assume a constant recovery rate when credit derivatives are valued. This is dangerous. When the default rate is high, the recovery rate for a particular asset class can be expected to decline. This is because a high default rate leads to more of the assets coming on the market and a reduction in their prices.[4]

As is now well known, this argument is particularly true for residential mortgages. In a normal market a recovery rate of about 75% is often assumed for this asset class. If this is assumed to be the recovery rate in all situations, the worst possible loss on a portfolio of residential mortgages given by the model would be 25%, and the 25% to 100% senior tranche of an ABS created from the mortgages could reasonably be assumed to be safe. In fact, recovery rates on mortgages have declined since 2007 and are now around 25%.

Tail correlation

The standard market model for evaluating structures such as those in figures 2 to 4 is the Gaussian copula model. This model is likely to be the starting point for a financial institution. However, the model has a serious weakness. It does not reflect a well-known feature of financial markets known as 'tail correlation'. This refers to the tendency for the correlation between two market variables to increase in extreme market conditions.

Tail correlation is a feature of credit markets. It explains why a relatively high correlation is used in the Gaussian copula model when the senior tranches of standard portfolios such as iTraxx and CDX are valued. These are tranches that provide payoffs only in extreme market conditions. One way of handling tail correlation is to increase the correlation when risks are being evaluated. Another better approach is to examine the effect of using a model that incorporates tail correlation. Hull and White (2004) propose the use of a 'double t' copula. Anderson and Sidenius (2004) propose a copula where factor loadings depend on factor values. Both approaches provide a good fit to market data in the synthetic CDO market and are natural choices for understanding risks in the cash market.

Hull and White (2009) examine how expected losses on AAA-rated tranches are affected when the double t-copula is substituted for the Gaussian copula. They find that the ABS CDO tranches, and to a lesser extent the ABS tranches, that were rated AAA become much riskier.

Stress testing

The risk measures used by regulators, and by financial institutions themselves, are largely based on historical experience. For example, Value at Risk measures for market risk are typically based on the movements in market variables seen over the last

---

4 The negative relationship between recovery rates and default rates has been documented for bonds by Altman et al (2005) and Moody's Investors Service (2008).

two to four years. Credit risk measures are based on default experience stretching back over 100 years. Stress testing often involves looking at the largest market moves experienced over the last 10, 20, or 30 years.

There can be no question that historical data provides a useful guide for risk managers. But historical data cannot be used in conjunction with models in a mechanistic way to determine if risks are acceptable. In risk management it is important that models be supplemented with managerial judgment. A risk management committee consisting of senior managers should meet regularly to consider the key risks facing a financial institution. Stress tests should be based on the scenarios generated by these managers in addition to those generated from historical data. The risk committee should be particularly sensitive to situations where the market appears to be showing bouts of 'irrational exuberance'.[5]

Involving senior management in stress testing is important for two reasons. First, senior management can draw on a wealth of experience to propose credible scenarios for stress testing. Second, key decision-makers within a financial institution are more likely to take the results of stress testing seriously if their senior management colleagues have had a hand in generating the scenarios. It appears that the results of stress testing were not taken seriously during the period leading up to July 2007. In 2005 and 2006, many economists were forecasting a decline in house prices in the US. Risk management groups did consider this possibility in their scenario analysis. But few financial institutions took this into account in their decision-making.

Liquidity and transparency

The management of liquidity risk is receiving more attention as a result of the credit crisis. The crisis emphasises the relationship between liquidity and transparency. ABSs and ABS CDOs are typically defined by a legal document several hundred pages long. They are extremely complex instruments. One structure frequently references many other structures. As mentioned earlier, many investors relied on credit ratings and did not analyse the structures for themselves. It is hardly surprising that, once the market lost confidence in sub-prime mortgages, it became impossible to trade ABSs and ABS CDOs at other than fire-sale prices. The key lesson from all this is that one way of managing liquidity is to focus on relatively simple structures where the cash flows in different future scenarios can be easily evaluated.

Collateral

Another lesson from the crisis is that all transactions, regardless of the credit ratings of the two sides to the transactions, should be collateralised. Regulators should insist on this. In some cases, the collateralisation will involve clearing houses such as those that have been developed for CDSs. In other cases it will involve two-way collaterali-

---

5   Irrational exuberance was a term coined by Alan Greenspan during the bull market of the 1990s.

sation agreements with zero thresholds. It will hopefully never again be possible for a financial institution such as AIG to create havoc in financial markets by taking a huge exposure without being required to post collateral.

Compensation

Compensation schemes are not the responsibility of risk management, but they have a big impact on the risks being taken by financial institutions.

Employee compensation falls into three categories: regular salary, the end-of-year bonus, and stock or stock options. Many employees at all levels of seniority in financial institutions, particularly traders, receive much of their compensation in the form of end-of-year bonuses. This form of compensation tends to focus the attention of the employee on short-term results.

If an employee generates huge profits one year and is responsible for severe losses the next year, the employee will receive a big bonus the first year and will not have to return it the following year. The employee might lose his or her job as a result of the second year losses, but even that is not a disaster. Financial institutions seem to be surprisingly willing to recruit individuals with losses on their resumes. Imagine you are an employee of a financial institution buying tranches of ABSs and ABS CDOs in 2006. Almost certainly you would have recognised that there was a bubble in the US housing market and would expect that bubble to burst sooner or later. However, it is possible that you would decide to continue trading. If the bubble did not burst until after 31 December 2006, you would still receive a bonus for the year ending on that date.

It is not necessarily the case that salaries on Wall Street are too high. Instead, it is the case that they should be calculated differently. It would be an improvement if annual bonuses reflected performance over a longer period of time than one year (say, five years). One idea is the following. At the end of each year a financial institution awards a 'bonus accrual' (positive or negative) to each employee reflecting the employee's contribution to the business. The actual cash bonus received by an employee at the end of a year would be the average bonus accrual over the previous five years or zero, whichever is higher. For the purpose of this calculation, bonus accruals would be set equal to zero for years prior to the employee joining the financial institution (unless the employee manages to negotiate otherwise) and bonuses would not be paid after an employee leaves it. Although not perfect, this type of plan would motivate employees to use a multi-year time horizon when making decisions.

Conclusions

Many factors contributed to the financial crisis that started in 2007. Mortgage originators used lax lending standards. Products were developed to enable mortgage originators to profitably transfer credit risk to investors. Rating agencies moved from their traditional business of rating bonds to rating structured products. The products bought by investors were complex and in many instances investors and rating agencies had inaccurate or incomplete information about the quality of the underlying assets.

There are a number of lessons for financial institutions and their risk management groups. Building models is important not only to value products but to understand the risks being taken and the dangers in the assumptions being made by traders. There is a tendency for Value at Risk calculations to assume that correlations observed in normal market conditions will apply in stressed market conditions. In practice, it is found that correlations almost always increase in stressed market conditions. Transparency is an important feature of a financial product. If the cash flows from a product cannot be calculated in a relatively straightforward way, the product should not be traded. Collateralisation should become a requirement in over-the-counter (OTC) markets. Compensation schemes should be chosen to better align the interests of traders with the long-term interest of the financial institutions they work for.

References

Altman, Edward I., Brooks Brady, Andrea Resti, and Andrea Sironi. 2005. 'The link between default and recovery rates: implications for credit risk models and procyclicality'. *Journal of Business*, 78, 6: 2203–2228.

Andersen, Leif and Jacob Sidenius. 2004. 'Extensions to the Gaussian Copula: Random Recovery and Random Factor Loadings'. *Journal of Credit Risk*, 1, 1: 29–70.

Gorton, Gary. 2008. 'The Panic of 2007'. Working Paper. Yale School of Management.

Hull, John C. 2008. 'The Financial Crisis of 2007: Another Case of Irrational Exuberance'. In *The Finance Crisis and Rescue: What Went Wrong? Why? What Lessons Can be Learned*, Toronto: University of Toronto Press.

Hull, John C. 2009a. 'Risk Management and Financial Institutions', 2nd Edition, Upper Saddle River, New Jersey: Pearson.

Hull, John C. 2009b. 'The Credit Crunch of 2007: What Went Wrong? Why? What Lessons Can Be Learned?' *Journal of Credit Risk*, forthcoming.

Hull, John. C. and Alan White. 2004. 'Valuation of a CDO and an nth to Default Swap Without Monte Carlo Simulation'. *Journal of Derivatives*, 12, 2: 8–23.

Hull, John C. and Alan White. 2009. 'The Risk of Tranches Created from Sub-prime Mortgages,' Working paper, University of Toronto.

Keys, Benjamin J., Tanmoy Mukherjee, Amit Seru, and Vikrant Vig. 2008. 'Did Securitisation Lead to Lax Screening? Evidence from Sub-prime Loans'. Working paper, University of Michigan.

Mian, Atif R. and Amir Sufi. 2008. 'The Consequences of Mortgage Credit Expansion: Evidence from the 2007 Mortgage Default Crisis'. Working Paper, Graduate School of Business, University of Chicago.

Moody's Investors Service. 2008. 'Corporate Default and Recovery Rates, 1920–2007'.

Zimmerman, Thomas. 2007. 'The Great Sub-prime Meltdown'. *The Journal of Structured Finance,* Fall: 7–20.

About the author

David Lando is Professor of Finance and Head of the Department of Finance, Copenhagen Business School. He holds a master's degree in mathematics-economics from the University of Copenhagen and a Ph.D. in statistics from Cornell University. His main areas of research are credit risk modelling and financial risk management. He has published several widely-cited papers on credit risk modelling and is the author of a monograph on the same topic published by Princeton University Press. Prior to joining Copenhagen Business School he was a professor at the University of Copenhagen. He has held visiting positions at Princeton University, The Federal Reserve Bank of New York and the Federal Reserve Board in Washington, and he has served as a member of Moody's Academic Advisory and Research Committee since its inception in 2001. Contact: dl.fi@cbs.dk

4
Some thoughts on the role of
mathematical models in light of the crisis

ABSTRACT The financial crisis once again has put focus on the role of mathematical models in financial markets. This paper discusses some of the dangers of mathematical models but also argues that we cannot do without them. We need models to communicate prices of similar instruments on a transformed scale but this need also brings pressure towards too much simplification and towards applying the models outside their domain. We still need models to help us pose sharp questions. A concrete example of the modelling of collateralised debt obligations serves to illustrate the general points.

4 Some thoughts on the role of mathematical models in the light of the crisis

by David Lando

Introduction

Bad mathematical models or too strong reliance on mathematical models is frequently cited as one of the culprits of the financial crisis. While models surely played a role, putting such a responsibility on models is giving too much credit to the influence models have in the larger strategic decisions in financial institutions and in political decision-making. Decisions to run banks with extreme leverage, to systematically try to loosen up capital constraints using off-balance sheet vehicles, searching-for-yield behaviour by institutional and private investors, political desires to increase house ownership even at the cost of increasing riskiness in lending – these decisions are not deeply rooted in models. When they are, one is suspicious that the model selection has often been such that the decisions were justified in the selected models.

This does not mean that we should not again discuss the role that models play in financial decision-making. The financial crisis adds new examples to frame this discussion and to help future decision-makers understand the trade-offs and the dangers involved in using models. Decision-makers must understand these dangers, for we cannot live without the models. Even if financial products become simpler and more transparent in the future and if the financial architecture becomes simpler, we depend critically on models. Below, I will try to articulate why this is so and I will use an example to discuss the models and the role they play by looking at three types of models used to model collateralised debt obligations (CDOs). To help readers for whom this is an unknown type of instrument, I will define the pay-off structure, somewhat idealised, of two such products and I will explain as non-technically as possible the basic model structure in three types of approaches.

Why models are useful

Kac (1969) contains one of the clearest statements of what mathematical modelling is all about:

"Models are for the most part, caricatures of reality, but if they are good, then like good caricatures they portray, perhaps in a distorted manner, some of the features of the real world. The main role of models is not so much to explain and to predict, though ultimately these are the main goals of science, as to polarise thinking and pose sharp questions."

To most people working as researchers using mathematical models, the statement is in a sense obvious. Models are indeed caricatures. But why only 'for the most part'? In physics there are a few natural laws whose empirical power and predictive ability make the models within which the laws are formulated seem like more than just good caricatures. The law of gravitation which states that the force of attraction between two bodies decreases with the square of the distance between them is an example of a law whose empirical validity has been established to an unbelievable degree of accuracy. We have become used to the power of this law allowing us to predict exactly when a comet can be seen in the sky and to send space ships to the moon. It is important to realise that there is simply no parallel to such a law in economics. In fact, even in physics, mathematical models are often caricatures with varying degrees of accuracy and predictive power. So the idea that mathematical models should be able to predict where prices are going and when a crisis will begin is hopeless – at least for the foreseeable future.

So why then are the models useful? As we will see below, Mark Kac's observations are still accurate and I will try to illustrate why, for example, they may help us ask the right questions. But I will also address the use of models in financial markets to communicate prices of similar instruments on a common scale. This role of models is perhaps the most important reason why models will continue to play an important role in financial markets. But it also contains one of the most important dangers which is that the demand for ease of communication leads markets to favour models that are too simplified.

The discussion quickly becomes too abstract without an example and we therefore look at CDO modelling to illustrate the main points.

The modelling of cash flow CDOs and synthetic CDOs – a non-technical introduction

To have an example to work from, this section provides a non-technical and idealised description of how some main types of CDOs work. Readers familiar with these types of contracts can skip this section.

Markets for CDOs had an explosive growth over the last decade. Much of the growth has been driven by regulation, but other motives for securitisation have been at play as well. The structure we describe below has for example allowed the originators to create securities with very different risk profiles from a relatively homogeneous set of financial assets thus catering to investors with different risk appetites.

To understand the basic structure, imagine that a bank has 100 loans each with a principal of €1m whose default risk the bank wishes to off-load. The bank cannot or may not wish to terminate the relationship with the borrowers, but consistent with its 'originate-to-distribute' philosophy it wants to free up regulatory capital to develop business with new clients. The bank creates a company – a special purpose vehicle (SPV) – that buys the loans from the bank. The loans form the collateral pool. The funding for this purchase is obtained by selling prioritised claims, so-called 'tranches', to the cash flows from the loans. As a very simple example, one could imagine that the loans are zero-coupon bonds and that they as well as the tranches mature in one year. Also, we may assume that here are three tranches: a senior tranche with principal of €60m, a junior tranche with principal of €30m, and an equity tranche with a principal of €10m. If the full principal of the loans is repaid, all of the tranches receive their full principal. But if losses occur in the collateral pool, the equity tranches are the first to be hit. All losses up to €10m in the underlying collateral pool are born by the holders of the equity tranche. If losses exceed €10m, the payment to holders of junior tranches is reduced with the amount by which the loss exceeds €10m up until the loss hits more than €40m, at which point the equity and junior tranches have lost everything and the senior tranche now begins to take losses. In practice, things are more complicated. The underlying loans as well as the tranches pay coupons. Complicated so-called waterfall structures describe the precise way in which coupons from the underlying loans and recovery from losses that occur before maturity are distributed to the tranches. The waterfall structure is meant to ensure that cash is not paid to lower priority tranches unless the more senior tranches are still 'safe'. If necessary, the principal of senior tranches is reduced through prepayment when the value of the collateral pool starts eating its way down to threatening levels.

In the structure described above, the protection sellers are the buyers of the tranches; if there are defaults, they lose money. Furthermore, it is a funded structure in which the buyers of the tranches have to make an upfront payment or 'funding', which finances the purchase of the collateral pool. We could also use the loans merely as reference securities in an 'unfunded' transaction. This would work, again in an idealised structure, as follows: The protection buyer now buys protection on losses within a certain range in the reference pool. To be more concrete, one could imagine buying protection against losses between 3% and 6% in the reference pool. The principal of the contract can be any amount. The price of the protection is a periodic premium but there is no cash paid at the initiation of the contract. The seller of protection then compensates the buyer when losses in the reference pool exceed 3%. The contract terminates at maturity or when losses exceed 6%. At that point the entire principal of the contract has been paid from the protection seller to the protection buyer.

Since the underlying loans are only serving as reference credits here, we can write contracts that have a much higher principal than the principal in the collateral pool. In fact, we do not need a principal in the collateral pool. We just need to keep track of whether there is a default and what the recovery is. An easy way to keep track of this is to use credit default swap (CDS) contracts as reference securities. This is what is done in the standardised synthetic CDO markets, which use names in the so-called CDX and Itraxx indices as reference securities. They are called standardised because the contract terms have become standardised and simplified, avoiding complex waterfall structures, and they are called synthetic because the reference securities are CDS contracts rather than actual bonds or loans.

Three classes of models

The basic challenge in modelling prices of CDO tranches is capturing the default probabilities of the underlying reference names, their recovery in default and the correlation between default events. It is of critical importance to the value of the different tranches in a CDO whether the reference names have a large risk of defaulting in clusters, i.e. if there is a larger probability of many defaults within a short period than there would be if default of the names could be viewed as independent events. It is not difficult to describe in a few equations how the main modelling approaches deal with the problem.

In model 1, we define the event of the default of an individual issuer as follows:

$$v_i \equiv \rho M + \sqrt{1 - \rho^2}\, \epsilon_i < K_i$$

This is consistent with an interpretation going back to a classical model of corporate bond pricing analysed in Merton (1974) where we view default as happening when the market value of a firm's assets falls below a certain threshold. We here assume that the logarithm of the assets is normally distributed and represent this logarithm as the sum of two shocks: one coming from the 'market', as represented by M and one coming from firm specific developments represented by ε_i. This approach was used in Vasicek (1991). The choice of the threshold is made so that it matches the probability of default inferred from premia on CDSs. This is typical of an arbitrage pricing framework in which we price more complicated claims to be consistent with pricing of the more primitive building blocks from which it is constructed.

The key parameter in the model is ρ. This parameter controls the extent to which defaults tend to occur in clusters. If firms have high values of ρ, their asset values fluctuate together going either up or down at the same time. When ρ is small, they do not tend to cluster. This model is static. It looks at the problem at one point in time. It tells us nothing about where we would expect the market to be tomorrow, and how our exposures change over time.

To state model 2, we need to define the notion of a stochastic intensity process. If a company has default intensity $\lambda(t)$ it means that given the firm has survived up to time t and given information I(t) on the state of the economy at time t, the probability of the firm defaulting over the next little instant Δt is given by:

$$P\big(\tau \leq t + \Delta t \,\big|\, \tau > t, I(t)\big) = \lambda(t)\Delta t$$

Here, τ denotes the default time of the firm. The relationship is actually only exact in the limit, but to give an idea of why intensities are useful, think of the following: if we model a distribution of the default time of a firm, and we imagine we can describe the distribution through a probability density function, then it is possible to represent the distribution of the default time using a so-called hazard function h as follows:

$$P(\tau > t) = \exp\left(-\int_0^t h(s)\,ds\right)$$

The hazard function can be viewed as an intensity when no information other than survival is taken into account, in other words we have:

$$P\big(\tau \leq t + \Delta t \,\big|\, \tau > t\big) = h(t)\Delta t$$

If we model the survival distribution through a stochastic intensity, we have that:

$$P(\tau > t) = E\exp\left(-\int_0^t \lambda(s)\,ds\right)$$

Hence, we can think of the intensity as giving us a 'random set of hazard rates' over which we average to find the default probability. The intensity, however, is a dynamic process whose value changes as we continuously learn about the state of the economy and the firm. The hazard function is a static function. For a more rigorous treatment of this framework, see Lando (2004). For short maturities, the premium on a CDS with the firm as reference credit is roughly equal to the product of the intensity and the loss in default. Hence if we are willing to say what the loss in default is, then we can infer default intensities from CDS spreads.

Now we can explain the structure of a second class of models: Model 2 specifies the default intensity of a firm as depending on two factors, one factor μ that influences intensities of all firms in the pool that we are modelling and one v^i which is specific to each firm. In short:

$$\lambda^i(t) = \alpha^i \mu(t) + v^i(t)$$

The sensitivity of firm i's intensity to changes in the common factor is given by a parameter α^i. This framework is from Mortensen (2006) using a small extension of Duffie and Garleanu (2001). The challenge in this framework is to model the stochastic evolution of the intensity. The systematic parts are diffusing round a long-run mean, but occasionally they take large jumps. The degree to which the large jumps happen simultaneously for all firms or whether they are highly firm specific is critical in determining the correlation structure. It becomes this choice that plays a role similar to the role of the correlation parameter ρ in Model 1. Note that model 2 is a fully dynamic model. Hence the parameters are determined not only from static 'calibration' requirements, i.e. requirement that the prices one computes on the various tranches fit at a specific point in time, but also from the time series behaviour of the intensities. This means that the model is rich enough to analyse questions of hedging, risk premia, pricing of so-called 'bespoke' tranches, i.e. tranches that are not standard because the loss limits defining the tranches or the CDS contracts used as reference securities are non-standard. One can also analyse what happens if one credit shoots up, if they all increase, etc. There is only one cost for this richness: there are many parameters to estimate. We essentially model the dynamic evolution of each firm's CDS premium and how they are correlated with other firms' CDS premia and this requires a lot of parameters.

As a final example, model 3 looks only at the aggregate loss of the pool directly as a single process. The simplest example of such a 'top-down' model is the model proposed by Longstaff and Rajan (2008). Let L(t) denote the aggregate loss in the reference pool by time t, i.e. if the collateral pool consists of 100 loans of size €1m in size, 10 loans have defaulted and the loss rate has been 0.6 on all loans, then the value of L(t) is €6m. Longstaff and Rajan (2008) assume that losses of different (but fixed) sizes hit the pool with varying intensities. Formally stated, model 3 has the following structure:

$$L(t) = 1 - \exp\left(-\gamma_1 N_{1t}\right) \exp\left(-\gamma_2 N_{2t}\right) \exp\left(-\gamma_3 N_{3t}\right)$$

Here, N_{it} it counts the number of losses of type I that has hit the pool by time t. When the γ_1 is small, it roughly represents the fraction of the remaining pool lost when a jump of type I occurs. Each of the counting processes N_i has a stochastic intensity process as defined in model 2 – controlling the frequency by which the jumps occur. For the precise specification see Longstaff and Rajan (2008). Note that whereas intensities were used in model 2 to represent the specific default intensity of each firm, intensities are now used to model sizes of losses to the aggregate pool. Since some of the losses in their model can be larger than the worst possible loss of one firm, we must interpret these events as simultaneous defaults of many firms in the collateral pool. Determining the jump intensities and the sizes of the different loss severities that give the best fit is a delicate problem. Longstaff and Rajan find that the assumptions that best fit a specific time interval of data on the CDXindex, corresponds to assuming three types of losses occurring with certain probabilities, under the probabilities used for pricing contracts, the so-called risk neutral probabilities: 40 bp loss on average every year, 6% loss on average every 40 years, and a 34.6% loss every 800 years. Note that in this kind of model there is no modelling of the underlying names. We are unable to distinguish between whether a high default risk name or a low default risk name leaves the pool. One also becomes suspicious that the fitting is very dependent on how the tranching is defined. In terms of parsimony, model 3 lands between the two models we considered already.

The three models above will be our benchmarks for the following discussion of models.

Models as communication devices

One of the most important functions of models in the market place is their role as communication devices. Prices of similar instruments that differ in some simple dimension, such as maturity, are not easy to compare, and we therefore benefit from transforming prices into a common scale. We are used to comparing bonds in term of their yields, since the yield is an easier quantity to compare than prices of bonds whose maturities and coupons differ. For mortgage bonds, the notion of option adjusted spread is a model-dependent way of representing that part of a mortgage bond's excess yield over a treasury yield which remains even when we correct for the options embedded in the callable mortgage bond. Transformations such as these are not unique to fixed-income markets. They are an integral part of derivatives markets as well.

The most prominent example of such a model-based transformation is the notion of implied volatility in options markets. In the Black-Scholes option pricing formula, the only input to the pricing formula that is not either directly observable or stipulated in the contract is the volatility of returns of the underlying security. In fact, one of the virtues of the Black-Scholes model is that it explains why volatility and not expected returns of the underlying security enters into the pricing of options. Implied volatility is simply the value of the volatility one should plug into the Black-Scholes formula so that the model price matches the observed option price in the market. If the Black-Scholes model were a physical law, options with different strike prices and different maturities trading on the same underlying asset, would have the same implied volatility. The fact that the implied volatility is not constant across strikes or maturities is proof that the Black-Scholes model does not hold. Modellers, and traders hopefully, know this. Yet the implied volatility still reveals how options of different strikes are priced compared to a Black-Scholes model and therefore relative to each other. The fact that there are 'smiles' in implied volatility curves, i.e. implied volatility for deep out-of-the-money puts is higher than for at-the-money puts, is a sign that investors in some sense are paying 'more' for these options, which in essence serve as insurance against steep losses. The smile is a robust empirical phenomenon across different types of options and different underlying instruments. Introducing model modifications such as jumps in the value of the underlying security or stochastic volatility delivers option prices which, if the model is true, will display smiles when transformed back to implied volatility using the Black-Scholes formula. This contributes to our understanding of why the market is deviating from the Black-Scholes model. The implied volatility is also a reasonable interpolation device. Even if markets do not follow the Black-Scholes model, the price of an option with a given strike is pretty close to the price obtained by interpolating implied volatilities from options with smaller and larger strikes.

When the market for CDOs began to grow, the search for a good pricing model was intense both in practitioner communities and among academics. In this process, practitioners call for analytic tractability and look for models that allow quick computation. This is an interesting challenge but it sometimes pushes to the background the search for models that allow for a deeper understanding of risk premia, sources of correlations and connections to other financial markets. An equally strong push towards simplicity comes from the market's desire to have a simple model that may serve as a communication device in the same spirit as the Black-Scholes model. In the context of CDOs, this led to the market adopting model 1 because it is capable of delivering one number which, under some fixed input assumptions, summarises the price of a tranche in a CDO. If default probabilities are taken as given and the correlation is assumed to be the same between all names in the reference pool, then the correlation parameter can be used to transform prices into a common scale. Just as implied volatility is not the same for different options even when the underlying security is the same, we also need different correlations to match different CDO tranches written on the same pool. This is again proof that the model is wrong. But perhaps, just as with Black-Scholes, this does not matter. The transformation might still be useful. Unfortunately, it was evident early on that, in addition to having problems fitting real data, the model had some conceptual difficulties which showed that it would not have the same virtues as implied volatility.

First of all, for mezzanine tranches, the implied correlation is in general not unique. This is a consequence of the 'call spread' nature of the pay-offs to these tranches – that they are in essence a difference of two options whose prices are both increasing in correlation. Secondly, it quickly became clear, that there were days when contracts traded, where the price of the tranches could not be inverted into an implied correlation. The observed prices were simply out of range of the pricing model. Third, the model was a static model, essentially only used to try to calibrate data at a given day. The dynamics of the model were not added and if they were, they were inconsistent with the specification of the mechanism that triggers default.

The non-uniqueness problem was addressed through a modification known as base correlations, which essentially convert the inversion problem into looking at equity tranches only. But even this technique did not resolve the issue of non-existence. Also for base correlations, there were days when the measure could not be inverted from prices. And there were some more conceptual problems laid out in Willemann (2005):

First, if we assume we know the true default dynamics of underling firms in a fully specified model similar to model 2, what will happen to base correlations when we increase the true correlation in the model? If the base correlation measure were indeed representing only correlation, it should increase as well. It does not necessarily do this. That is, we cannot know if an increase in base correlation is due to increasing correlation in the underlying names. Second, base correlation permits negative prices of non-standard tranches. That is, when we try to use the measure as an interpolation device and price non-standard tranches, i.e. custom-made tranches for which there were no market quotes, it sometimes produces negative prices of contracts whose cash flows are always non-negative. There are other, more subtle, problems with base correlations that are addressed in Willemann (2005).

The inability of the model to really interpret what was happening in markets became evident already during the 'correlation' crisis, which hit in May 2005. The crisis is described in several papers, including Kherraz (2006). Without going into much detail, the crisis was triggered by downgrades of Ford and GM by the major rating agencies, and this caused large unwinding of positions which combined selling of equity protection with buying mezzanine protection, a trading strategy which suffered large losses as the unwinding of the positions caused losses both on the short and the long side of the deal. The failed hedge was a demonstration that the hedging based on the model surely did not work. There was no room in the model to allow for changes in the correlation. Attempts to fix the model, by altering the distributional assumptions gave better fits at any instant in time at the cost of using more parameters but it did not improve our understanding of the dynamic behaviour of the markets and of the risk management implications.

The correlation crisis was a warning that the models for CDOs did not work. This was true even for standardised products that are easier to work with since there are good alternative sources of information on the underlying names from equity markets, corporate bond markets and CDS markets. The interesting part is that the apparent lack of a good model did not prevent heavy trading in the products. One could have hoped that the intellectually more satisfying way to model prices of CDO tranches with a full dynamic model from the bottom up as in model 2 above would have fared better. But that would only potentially have worked if the model had included the possibility of changing correlations between the underlying CDS names. In today's markets, model 2 has problems as well. One key problem is that it is hard for the model to capture the extreme volatility in the senior tranche premia. So even if the model is much better at addressing the mechanics that drive the different tranche prices, it still has problems capturing the current market conditions. Hence the fact that it is hard to use as a simple communication device is not the only problem with model 2. A plausible explanation for the fitting problems is that shortage of capital constrains potential sellers of protection from bringing prices back to 'resaonable levels'.

Model 3 based on aggregate loss offers no solution. This model has many of the same shortcomings. How are we able to understand the changes in the dynamics of the loss process without understanding the components? That is, if we are looking at a market where we are trying to use the model to price contracts and not just calibrate to explain observed market prices in a liquid market, then we cannot avoid modelling the individual components. This is the biggest limitation of the top-down models. They are aimed at pricing derivatives on tranches but this market has more or less died out.

The danger of oversimplification and overextending the domain

There is a much more serious consequence of abolishing models that work from the bottom up. When the CDOs become very complicated because the collateral pool contains mortgage bonds with extremely complicated characteristics or tranches of other CDOs, then model 1 becomes an oversimplification. As a minimum we have to address that the mixtures of products underlying the structures have vastly different correlations, and when instruments in the collateral pool become tranches of other CDOs, it becomes very hard to attribute meaning to the correlation parameter in this model. In standardised CDOs as described above, we have the benefit of having more information on each name in the collateral pool. We can check if the correlations inputs are matched by correlations observed for the same entities in other markets. When the underlying tranches in asset backed securities (ABS) CDOs are tranches of other CDOs or mortgage bonds mixed together, the correlation becomes very hard to verify or even to interpret. If we do not model these complex securities from the bottom up, the extreme correlation that may occur when junior tranches are used as collateral in CDOs is impossible to capture.

As an aside, the complexity of some traded derivatives contracts has always been a bit of a mystery to academics working in asset pricing. Who are the investors demanding the most exotic products? Surely, the financial institution structuring the deal in a derivatives contract has superior access to trading and hedging the contracts and essentially eliminates its side of the risk exposure, i.e. the institutions can engineer the pay-offs much less expensively than their clients. But with the highly exotic products, the clients have a very hard time judging whether the pricing is reasonable. Many derivatives are so complex that there are no simple pricing formulas. How can the investor judge the fairness of the price? Investors should be aware that when derivatives are very complex, financial institutions are taking compensation for the significant 'model risk' that arises from trading the contracts.

Interpreting signals and warnings in the market

In addition to their role as a tool for communication and for making comparisons, models are needed if we want to meaningfully assess what market prices are telling us. The market did not predict the crisis. Any price is an average over favourable and non-favourable outcomes, and ex post when the bad outcome has occurred it is easy to blame the model. But it is worth noting that both the Longstaff and Rajan (2008) paper and Mortensen (2006) in their calibration to market prices inferred that relatively extreme scenarios were priced in. We have already seen the probabilities of extreme losses reported in Longstaff and Rajan (2008) but Mortensen also showed that in quiet times jumps in CDS spreads of several hundred basis points were needed to calibrate to market prices. Of course, these probabilities are risk adjusted and not representative of real world probabilities, but they still force us to either think of extreme outcomes as very possible or, if we refuse to believe this, to have an opinion on why the market is so risk averse that it prices using much larger risk-adjusted probabilities of default than actual probabilities. Without models we can't even begin to ask these questions.

The choice of scenarios

It is important to realise that no model available could predict the spread movements we saw. Thinking about the models as predictors is confusing models with laws. As an illustration, consider the developments in the premium on different tranches in the standardised CDOs trading on names in the CDX index known as North American Investment Grade. This index consists of 125 investment grade companies across different sectors chosen on the basis of the liquidity of the CDS contracts trading on the companies. The development in these premia is shown in figure 1. For example, premia on the 15–30% tranche went from a minimum around 2 basis points up to a maximum of more than 120 basis points. The premium on the 3–7% tranche went from around 60 basis point to almost 11%. Note the very quiet period before the sub-prime crisis.

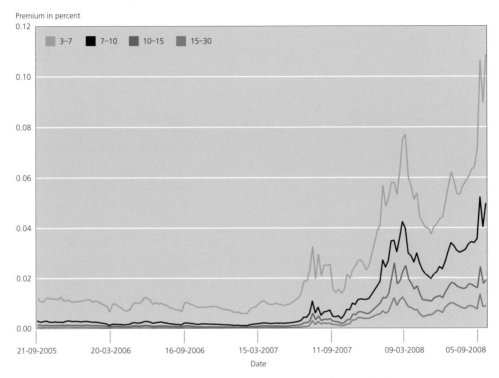

Fig. 1. CDX premiums for non-equity tranches. The development of premia for buying protection on different non-equity tranches on the North American Investment Grade companies in the so-called CDX index. The underlying pool consists of 125 investment grade companies in different sectors chosen according to the liquidity of the CDS contracts trading with the companies as reference credit. As an example, a premium of 1% on the senior 15–30% tranche means that a protection buyer must pay an annual premium of 1% (divided into four quarterly payments) of the insured notional. The protection seller only has to pay when losses exceed 15% on the CDS contracts. If recovery is 40% in default, this means that more than 25% of the loans have to default within 5 years for there to be any payment from the protection seller.

It is possible to include such extreme scenarios as possible scenarios in models. Ultimately, it is management and risk management who decides how much model inputs are to be stressed. Nothing prevents them from being conservative in the model inputs. Management must then decide whether capital should be put aside to withstand these extreme scenarios and subsequently explain to shareholders that this will cause expected returns to fall because of lower risk. Very few had the nerve or the power to do this in the quiet period.

The ability to pose sharp questions

Understanding how asset prices evolve in financial markets is a frustratingly compli-cated task and even if we have some rich theoretical frameworks in portfolio theory and in arbitrage pricing, there are still many phenomena we do not understand. Still, there are certain insights that we believe are true. Carrying systematic risk carries a risk premium. We can disagree on how we measure systematic risk, but carrying a risk that is more closely correlated with a pervasive economic shock should be re-warded differently from carrying a highly idiosyncratic, firm specific risk.

The disagreements in the finance profession on how to quantify systematic risk and the inability to find a standard way of measuring it, should not serve as an excuse for investors not to pose sharp questions based on the premise that systematic risk should be priced. A consequence of this is that if two risks are presented as compa-rable in terms of actual probabilities of occurring, it does not mean that the risks are necessarily priced in the same way. If a deal is presented as favorable because it offers high return compared to other claims with similar risks measured in actual prob-abilities, then the investor should take a closer look at the nature of the risk. This is not to say that favourable deals cannot exist, but long persistent mispricings are rare. The extra return typically comes at a price. The informed investor should try hard to understand which price is paid in terms of risk-taking for any extra reward.

A selling point of CDO tranches was often that they were structured so that the yield on the tranche was high compared to the rating. It was not uncommon to of-fer AA-rated tranches yielding LIBOR plus several hundred basis points. Recall that LIBOR before the sub-prime crisis was an interbank rate offered by bank, close to AA-ratings. Even if we ignore the fact that models were probably underestimating the probability of a severe downturn in house prices, there are still important questions that could have been asked with the benefit of a good model.

Investors should realise that ratings speak about the actual default probability whereas yields reflect both actual default probability and a compensation for bearing risk. The more systematic risk, the higher the premium should be. Using simple mod-els, one can formulate clearly the sense in which systematic risk is being concentrated in tranches. It is possible to show, as for example in Brennan, Hein and Poon (2009), that if some investors are willing to price CDO tranches of a given rating at the same level as a bond with similar rating, then it is possible to arbitrage the market for an arranger who buys the underlying bonds and sells tranched claims to the bonds' cash flows to these unsophisticated investors. As shown in that paper, such investor behav-iour was not uncommon. Of course, investors did get more yield than the underlying bond, but the selling argument that this is an 'arbitrage' does not hold. Insights from fundamental finance theory are helpful here in posing the sharp questions. We will need more sharp questions after the crisis. There are numerous other questions that arise after the crisis. In relation to CDOs, we will have to re-examine why pooling of securities and especially tranching of securities occur and under what circumstances they are a useful innovation. And we will have to be better at modelling the risk in these products and to structure them with so much transparency that we can do so.

Models as tools for structuring our data collection efforts

Good models structure the way we collect and interpret data. Even models that are caricatures are helpful in this respect if they do not assume away too much complexity. If models become too 'reduced-form' and in reality do not require the user to understand more fundamental dynamics of the model or the inputs to the model, then management has to be extra careful in relying too strongly on the conclusions. There is much black-box modelling out there. If one insists on modelling the pay-offs to complicated CDOs from the bottom up, then one is forced to address the fine structure of the underlying pay-offs. There is no way around such efforts. For even if a modeller argues that a simplified model can do the job, then to confirm that this simplified model is not an oversimplification, it will have to prove its robustness to variations in the underlying sources of data and that it in some sense performs almost as well as the more complicated model.

But the crisis does contain lessons for our modelling efforts. We will have to look more into measures of available capital, of concentrations of holdings, of leverage and of liquidity. There is a challenge in developing good operational models, beyond the conceptual clearing, which will help us focus on what data we should collect.

Conclusions

Mathematical models of CDOs did not cause the crisis. Attempts to blame the crisis on these models overestimate the influence of quantitative modelling in the financial industry in shaping decisions on capital structure of the banks, the choice of leverage, the systematic search for capital relief through creative accounting, and uncritical investors searching for yield. This is not to say that there have not been plenty of misguided trading and investing where agents have relied too much on models.

Models for CDOs were known to be a problem – the correlation crisis had already given plenty of warnings to that effect. The models of the rating agencies seemingly did not put enough weight on states with massive house price depreciation and did not take into account the poor documentation of sub-prime loans or account for the possibility of extreme correlation patterns in structured finance CDOs. But the losses that came from these mistakes could easily have been absorbed by a less levered financial system. Once again, a huge problem was leverage.

The crisis is an opportunity, with greater credibility perhaps, to repeat some warnings about models. Most importantly, mathematical models of financial markets, of economic systems in general, and indeed of almost any other area in which they are applied, are not laws. We cannot predict prices in financial markets with great accuracy. The changing nature of financial markets, their institutions and the rate of financial innovation, and the influence from events which truly fall outside our modelling territory, such as acts of war or terrorism, mean that prediction can never reach the level known from simple physical systems, such as planets moving around the sun. Macroeconomic models have become better at understanding some of the workings of the economy, but it is striking, and we must admit frustrating, that there is remarkably small consensus on even some of the most basic issues, such as the effects of fiscal policy, in macroeconomics as well.

Models can set up scenarios and we can use models to better understand the consequences of these scenarios, perhaps not down to the last penny, but when the models are good within the right order of magnitude. The scenarios we choose to include are grounded in history and experience but they are not the outcome of a model. Management must decide whether a scenario that involves a repeat of the Great Depression should be a real possibility in the model, i.e. an event with non-vanishing probability. Management must decide whether capital should be set aside to withstand such an event. If that choice is made, shareholders must understand that this causes the stock to be less risky thus sacrificing expected return.

If management decides that a return on equity should be very high, it inevitably involves higher risk taking. Higher returns are typically connected with higher leverage, and higher leverage means higher risk. In boom times there will always be a tendency to disregard the disasters of the past arguing that 'we fixed the problem that generated the previous crisis'.

Models help us in understanding trade-offs in decision-making but whenever we have a model, we should beware of stretching the model too far outside its domain.

Watch out for 'hyping' of models. The Gaussian copula model – as model 1 is sometimes called – is an incredibly simple and crude model. The use of the word copula makes it sound more complicated than it is. It is really the old approach from Vasicek (1991) and the basic structure is simple to explain.

One should not be intimidated by models. If a quant or a risk manager cannot explain a model in simple terms, it is often a problem with the quant. Management and boards needs people who are not intimidated by complicated models and with enough training in modelling that they know when the problem is truly a communication problem on the part of the modeller.

Bad decision-making leading up to the crisis comes from ignoring economic insights that can best be articulated or learned from models. If every investor understands that systematic risk is concentrated in CDO tranches, then that investor will ask more critical questions on the reasons why high rated tranches seem to offer such high yields. And if the systematic risk cannot explain it, then the investor must ask whether the rating is wrong or whether the underlying cash flows are priced incorrectly. The economic insights which are sharpened through our modelling will help us better analyse whether there is a future for asset securitisations involving both pooling and tranching, whether we should have a central clearing place for CDS contracts and how we should set capital requirements.

In fact, there were examples, even before the crisis, where the models did provide interesting answers. The issue of why the useful results are not penetrating to lawmakers and policy-makers is another question too complicated for this essay.

References

Brennan, Michael, Julia Hein and Ser-Huang Poon. 2009. 'Tranching and Rating'. Working paper. UCLA.

Duffie, Darrell and Nicolae Garleanu. 2001. 'Risk and Valuation of Collateralised Debt Obligations'. *Financial Analysts Journal.* 57(1): 41–59.

Kac, Mark. 1969. 'Some mathematical models in science'. *Science.* 166(3906): 695–699.

Kherraz, Ammar. 2006. 'The May 2005 correlation crisis: Did the models really fail?' Working paper. Imperial College London.

Lando, David. 2004. 'Credit risk modelling – theory and applications'. Princeton University Press.

Longstaff, Francis and Arvind Rajan. 2008. 'An empirical analysis of the pricing of Collateralised Debt Obligations'. *Journal of Finance.* 63(2): 529–562.

Merton, Robert. 1974. 'On the pricing of corporate debt: the risk structure of interest rates'. *Journal of Finance.* 29: 449–470.

Mortensen, Allan. 2006. 'Semi-analytic valuation of basket credit derivatives in intensity-based models'. *Journal of Derivatives.* 13(4): 8–26.

Vasicek, Oldrich. 1991. 'Limiting loan loss probability distributions'. Working paper. KMV Corporation.

Willemann, Søren. 2005. 'An evaluation of the base correlation framework for synthetic CDOs'. *Journal of Credit Risk* 1(4): 180–190.

About the author

Marc Schröter, M.Sc. EE and BA in Finance, is Vice President and Head of the Strategic Research department at SimCorp, for whom he has worked since 1995. He has consulted with a wide range of SimCorp clients working with buy-side investment managers across front, middle and back offices. Previously Head of Professional Services at SimCorp, since 2006 he has led Strategic Research at the firm where his team of international experts sets the strategic direction for the SimCorp Dimension integrated investment management system.

Contact: marc.schroeter@simcorp.com

5
Enterprise architectures for investment managers from a risk management perspective

ABSTRACT The financial crisis of 2008–2009 has highlighted a number of issues that financial institutions will focus on going forward. One of the most dominant business drivers in that context is an increased focus on enterprise risk management, that is, establishing a risk management framework that takes into account all risk types that the financial institution faces. Derived from this, it is expected that there will be more scrutiny from legislators who will also demand more detailed information and transparency. Likewise, investors who have lost money during the crisis will demand increased transparency and reporting to ensure confidence in their investments.

In order to meet these demands, financial institutions face a number of challenges related to their IT solutions. Consolidation of data across asset classes and systems to get a full overview of risk exposure and transparency in valuation of complex derivatives are two of the challenges. In order to reduce costs, some financial institutions might consider outsourcing solutions. However, depending on the architecture, this may impose more challenges with regards to data consolidation, timely access to information etc.

This article identifies the major current business drivers and the IT related strategic challenges financial institutions are facing in order to address these business drivers. A number of typical enterprise architectures and different ways of implementing them are discussed and it is analysed how each architecture is solving the strategic challenges. It is suggested that no matter which operating model a financial institution has chosen – including outsourcing – it should choose an enterprise architecture that enables a strong core to handle all enterprise data and main workflows.

5 Enterprise architectures for investment managers from a risk management perspective

by Marc Schröter

Introduction

The impact of the financial crisis of 2008–09 has been severe in the financial industry. Firms such as Lehman Brothers and Bear Stearns have disappeared and most financial institutions around the world have been affected by the crisis. Investors have lost billions on the collapse of these institutions and the associated decrease in asset prices in different markets. These events have really brought risk management on top of every investor's agenda and means that risk types such as counterparty and credit risk as well as systemic and liquidity risk now get equally much attention as market risk and operational risk previously have had.

Governments in US and Europe have provided subsidies to the financial industry and some firms are now under more or less governmental influence or control. At the time of writing the US Federal Reserve System has carried out a stress test, its so-called Supervisory Capital Assessment Program on 19 banks and the result showed that ten of the banks would require additional capital. While the stock markets in the short-term perceived it as good news that nine banks were solid, it can on the other hand also be perceived from the angle that more than half of the banks will not be long-term sustainable unless provided with additional capital injection.[1] As a result of the government subsidies many financial institutions now have a degree of public ownership. This, and also the fact that taxpayers have indirectly invested billions of euro's and dollars into the financial services industry, means that governments, clients and the public are increasingly sceptical towards the whole industry and are requesting insight and transparency.

It is a general point of view that one of the factors for leveraging the crisis was inabilities to understand the risks involved with some of the complex derivatives such as credit default swaps or collateralised mortgage obligations.[2] Whereas on a short-term basis, there may be a trend towards going back to basics by investing in more traditional asset classes often in the form of beta/index tracking, over the medium to longer term there will still be a pressure to outperform benchmarks, in other words a search for alpha.[3] The search for alpha may come from investment in hedge funds or hedge-fund-like strategies, structured products and derivatives as well as increased

---

1 Press release by US Federal Reserve (http://www.federalreserve.gov/newsevents/press/bcreg/20090507a.htm).
2 Wall Street & Technology Blog 14 October 2008: One Way to De-Mystify Derivatives Valuation.
3 Dushyant Shahrawat and Dayle Scher, TowerGroup: 2009 Top 10 Business Drivers, Strategic Responses, and IT Initiatives in Investment Management.

investment in global equities and alternative investments, such as real estate and private equity funds. This means that there will still be investment in derivatives and other 'difficult-to-price' products which will drive a need to increase transparency into these products by decomposing risk and pricing of securities that are not traded in liquid markets.

As a consequence of the market events in 2008–2009, there is pressure both from regulators and clients for insight into how financial institutions manage risk. Increased scrutiny, shorter deadlines and new demands from all parties are expected, and financial institutions will need to make their risk levels and risk management process transparent to regulators, clients and the public. Some may focus especially on this in order to gain competitive advantage. As examples of important recent legislation to take into account are the Markets in Financial Services Directive (MiFID) in Europe and Regulation National Market System (Reg NMS) in the United States which over the past years have been driving markets into more fragmentation, that is to say a given security can be traded on multiple execution venues. The main rationale of the legislation was to increase competition by allowing trading on other venues than the traditional exchanges. The result of this has been an introduction of numerous multilateral trading facilities (MTF) which in turn makes the search for liquidity more complex since there is no longer just one marketplace for a given stock. For financial institutions this will impose a challenge because 'the search for liquidity' is getting more complex. The number of counterparties and markets increases and at the same time legislators are demanding more transparency such as in terms of best execution documentation towards clients.

The financial crisis has led to significant pressure on revenues from falling assets under management (AuM). The pressure on revenues will force companies to focus on managing costs by improving efficiency or direct cost cutting. Consolidation or outsourcing will be a natural consideration and may impact the overall operational risk since it will change the financial institution's operating model and associated enterprise architecture.

Another aspect of the crisis is significant changes in the service providers' industry. As a consequence of this, the buy-side institutions will reassess transparency and risk of business models where they rely on service providers such as brokers for trade execution, derivatives structuring and valuation. They will also go over their counterparty risk towards these institutions. Hedge funds are expected to use multiple prime brokers. This, and an increased pressure for independent valuation of fund net asset value (NAV), will increase the role of fund administrators who manage funds on behalf of fund managers across custodians and prime brokers. Another example is the securities finance area, where some buy-side institutions may consider to take on lending and collateral management in-house instead for relying on third party service providers (TPSPs) with associated risks involved.

This article will focus on three prevailing business drivers in the current market. The drivers are 'risk management', 'reporting and disclosure' and 'compliance with legislation'. These business drivers impose some strategic challenges that market participants must address. Since IT is an imperative part of the operation of financial institutions, many of these challenges will be IT related in terms of architecture, data and applications (today, rating agencies such as Fitch include technology as one of their five rating categories which shows the importance of IT in any financial institution).[4] The article will discuss IT related challenges when addressing the business drivers and will compare some typical IT solutions/architectures to discuss how they solve these challenges. Many of the aspects and conclusions are valid for the financial industry as a whole, but the focus will be on buy-side investment managers.

Fig. 1. The link between business driver, strategic challenge and IT solution

Risk management, reporting and compliance with legislation

As discussed in the introduction, one of the most important current main business drivers is generally considered to be an increased focus on risk management.[5] Financial institutions will move from a focus on individual risk types to a more holistic approach where all risk types across the institution are taking into account in a common framework.[6,7,8,9,10] As a result of the recent market events, regulators are expected to react with new legislation and therefore compliance to impending legislation is considered another important business driver.[5] Finally, because of increased scrutiny of legislators, government ownership and interest in financial institutions as well as investor's increased demand for transparency and control, the final major business driver is reporting and disclosure.[5]

4   Richard Willsher: 'Credit ratings: IT in the spotlight', Journal of Applied IT and Investment Management April 2009.
5   Dushyant Shahrawat and Dayle Scher, TowerGroup: 2009 Top 10 Business Drivers, Strategic Responses, and IT Initiatives in Investment Management.
6   Stephen Bruel, TowerGroup: Understanding Securities Industry Risk by Using the TowerGroup Risk Management Matrix.
7   Ernst & Young: Managing Information Technology Risk. A global survey for the financial services industry.
8   SimCorp: Global Investment Management Risk Survey 2009.
9   Gartner: Key Issues and Research Agenda for Banking and Investment Services 2009.
10 The 2009 Ernst & Young business risk report. The top 10 risks for global business.

Holistic perspective on risk management

Within risk management, a number of concepts and frameworks such as governance, risk and compliance (GRC) and enterprise risk management (ERM) are getting more and more attention from financial institutions.[7,11] Some of the major legal initiatives in Europe and US are addressing exactly these aspects. Sarbanes-Oxley (SOX) is much about the governance perspective and the requirement for having procedures and policies in place. Basel II for banks, Solvency II for insurance and UCITS III and IV for investment funds define requirements both for quantitative measurement of risk across multiple risk types, compliance rules as well as governance and disclosure of information.

Governance is about definition of mechanisms to ensure that a company has procedures and policies for measuring, monitoring, following up and taking corrective actions to guarantee compliance. It is generally the responsibility of senior management to ensure proper governance of the risks of the company and it is typically implemented by establishing policies and procedures that are governed by various committees. Another aspect of governance is to ensure transparency of the policies and procedures both within and outside the organisation.

Risk management is about defining measures and tolerances of risk levels and using these actively to meet business objectives. Basel and Solvency address in their 'Pillar one' specifically credit risk, market risk and operational risk. 'Pillar two' provides a framework for addressing other risk types such as systemic risks, reputation risks, liquidity risk and so forth. Many risk types, as well as the underlying risk factors, are correlated so they need to be addressed in a common framework. The concept of considering all risks types across the organisation is called enterprise risk management (ERM) and means that financial institutions must identify all risks they are facing, quantify them and look at them in an enterprise perspective in order to take corrective action on breaches. Typically, the most common risks types are divided into the categories of market risk, credit risk, operational risk and other risks (such as systemic risk, reputational risk and others).[6,12]

Finally, compliance is the activity that ensures that processes, policies and controls are followed in order to comply with defined risk levels for legislation, industry/client restrictions as well as internal guidelines. Compliance control in financial institutions covers both post-trade and pre-trade compliance. Post-trade compliance is a validation of compliance against restrictions at the end of the day to document and manage breaches and also to discover breaches originating from market movements. Pre-trade compliance on the other hand is about discovering a potential breach before an order is sent to the market and auditing overrides by portfolio managers or traders.

---

11   The Compliance Consortium: Governance, Risk Management and Compliance: An Operational Approach.

12   Balin: Basel I, Basel II, and Emerging Markets: A Nontechnical Analysis, The Johns Hopkins University School of Advanced International Studies (SAIS), Washington DC 20036, USA.

Reporting and disclosure

Clients, regulators and the public in general will increase pressure for getting more information about investment managers' investment processes and risk levels. This will drive a need for having solid reporting infrastructures in place to be able to meet these requests. Investment managers may themselves want to put effort into increased disclosure to use it in their marketing and perhaps gain a competitive advantage. Documentation of risk management processes, risk exposure of investments and risk-adjusted performance will be some of the things investment managers will seek to report and disclose. Failure to do so may increase reputational risk.

Compliance with regulation

It is expected that there will be changes in legislation to address the problems that led to the financial crisis and to ensure that financial institutions do not take unacceptable risk. There is likely to be more scrutiny from regulators and they are likely to ask for more details with shorter deadlines for providing the answers. At the same time there will also be pressure from other bodies such as industry organisations and market practice groups. Even though not coming from regulators, investment managers will still seek to comply with industry standards and common agreed practices in order to protect reputation and remain competitive.

Fig. 2. Three major business drivers in the current financial market conditions

Strategic challenges

Investment managers are considering their strategic challenges to deal with each of the three major business drivers. There are many types of challenges to consider for investment managers, this section focuses primarily on the IT related strategic challenges.

Risk management

Within risk management there are a number of IT related strategic challenges:

Framework for enterprise risk management
The first challenge is to establish a framework for addressing all risk types from an enterprise perspective. The enterprise view of risk should address all the common risk

types, typically categorised into market risk, credit risk, operational risk and other types of risk.[6,12] Risk management on the one hand is about having systems for calculating risk according to various risk models and stress test capabilities. At the same time, however, it is also about more basic, but equally important IT related issues, such as consolidating data across systems.[13,14]

Market risk is the risk associated with a move in market factors like equity prices, interest rates, currency rates and so forth. Risk types such as liquidity risk and settlement risk are also typically categorised as market risk. Liquidity risk is the risk that an asset cannot be traded quickly enough in the market thereby causing a loss or a funding problem. Settlement risk is the risk that a counterparty does not deliver a security according to agreement, which in turn can lead to a funding problem.

Credit risk is the risk associated with a debtor's inability to re-pay an obligation. Risk types within this category also include counterparty risk which is the risk that a counterparty to a transaction is unable to make future payments. Other risk types such as sovereign risk which is the risk that a state nationalises a company or similar thereby decreasing the value, are also categorised as credit risk.

Operational risk is the risk associated with the failure of internal procedures and controls, errors by humans and systems or external events. By this definition, operational risk can originate from many different areas such as IT failure, fraud or error.

Market risk, credit risk and operational risk are the most common risk types in an enterprise framework, but there are others that should be considered. One of these is systemic risk which is the risk associated with the collapse of an entire market as opposed to single components, such as securities or counterparties, within that market. Especially it refers to the dependencies in a market where one event may lead to the collapse of the whole market. The sequence of events in the credit crisis starting with the collapse of the US sub-prime market, downgrades by the rating agencies and the potential problems in the credit default swaps market are good examples of this type of risk. After 2008, this type of risk will most likely receive increased attention by financial institutions. Other risk types are reputational risk which goes across many of the other risk types as well as legislation risk which is the risk of failing to meet legislation.

Quantification of risk

Having defined the overall enterprise risk framework, the second challenge is to quantify all risk types. In the past, there has been special focus on market risk, where the Value at Risk (VaR) methodology is the most common approach.

Market risk is associated with moves in market factors and since the market factors are often correlated, and results therefore not additive, it requires a consolidated

---

13   Marc Schröter, SimCorp: 'Outsourcing and Risk Management, time to redefine requirements?' *Journal of Applied IT and Investment Management*, vol 1, no 1, 2009.
14   Marc Schröter, SimCorp: 'White paper on Extended Enterprise Data Management', 2009.

approach where all positions and risk factors are taken into account at the same time. Basel II, Solvency II, UCITS III/IV as well as local financial services authority (FSA) requirements are in general based on this approach. For credit risk, Basel II suggests three different approaches, all based on some kind of weighting of exposure according to credit rating of the issuer. Other key ratios such as loss given default (LGD) or the Creditmetrics methodology can be used. Basel suggests a quantification method for operational risk and systemic risk may be measured by using stress test scenarios. In summary, there are standardised quantification methods for some risk types, but not for all. So the challenge for investment managers is partly to comply with standards and regulations but also to define their own quantification methods where there is no standard.

Valuation

A part of quantification of risks is to establish the correct value of a given security or derivatives contract. OTC derivatives and structured products may be complex and difficult to assess in terms of understanding the risk involved with the investment. Some buy-side investment managers have used external valuation services, for example supplied by their brokers, while others use their custodians or third party administrator (TPA) for this.[3] During the financial crisis the challenge of correct valuation was not only restricted to complex derivatives but was equally difficult for securities that during the crisis became illiquid, for example sub-prime loans and collateralised mortgage obligations. This area remains a challenge and there is still debate, also among regulators, as to how to overcome the issues, for example by moving from mark-to-market valuation to booked value or other methods.[15] Another related challenge in valuation and risk assessment is the need to correctly decompose derivatives and structured products to their main components in order to decompose their risk and show attribution from the fundamental risk factors.

This leads to the next challenge regarding risk management.

Transparency

Because some investment managers have relied on their brokers to do the valuation and others have entirely outsourced their valuation and risk calculation to companies providing this as a service, there may not be very good transparency. This concerns both the valuation models used as well as the whole process. Outsourcing valuation may be a good solution for some but in the end it is the investment manager who is responsible for valuation and risk management. Some investment managers are therefore likely to choose systems and processes that provide greater transparency in order to avoid black-box calculations where they are not able to account for the results or able to identify and correct errors in calculations.

---

15   Peter Needleman, Towers Perrin: 'The Last Word: The Financial Crisis: Is mark-to-market to blame?'

Data quality and enterprise data management

Another challenge is to ensure good data quality which is a precondition for all valuation and risk management. An error in a price or a yield used for theoretical valuation can ruin any risk assessment. This subject has been the subject of increased attention during the past years and is now one of the key priorities for many investment managers.[16,17] Trustworthy security, reference and market data are the precondition for all valuations and risk calculations. For example, if a stock split is missed or an equity re-classification is not registered, for instance from large cap to small cap, the performance and risk attribution calculations are wrong. Or imagine if different systems use different security or counterparty identifiers, then consolidation of information for risk management becomes impossible.

This is often a problem in the case of security classifications, but probably even more essential in the case of reference data such as counterparty information. If an investment manager has invested in a counterparty's stock, the counterparty's issued bonds, has unsettled trades with the counterparty and has exchanged collateral, then calculation of the total exposure towards the counterparty depends on the fact that it is actually possible to identify that it is the same counterparty.

In case of global operations with subsidiaries in different countries, for both the investment manager and the counterparty, it can be very difficult to ensure data quality that enables an enterprise-wide view of the risk and other key ratios. In order to address this challenge, the concept of enterprise data management (EDM) has been widely promoted during the last years and is getting increased attention from investment managers. The challenge is to ensure that all systems and calculations use the same basic core data. Along with good quality security, reference and market data, the investment manager must also ensure good quality in positions data. This means that it should be possible to rely on the security holdings, derivatives contracts and cash balances used in risk management and for making investment decisions. Positions should therefore at all times be reconciled against custodians and counterparties. This is a basic precondition but there are some challenges to be taken into account such as keeping positions updated with corporate actions for equities, for equity derivatives, holdings on loan, and positions given as collateral.

Consolidation of information

The next challenge is to make sure that it is possible to consolidate data for all the risk types covered by the enterprise risk management. Market risk is calculated based on risk factors which are correlated across asset classes. So a calculation should include positions within all asset classes in the same calculation. Similarly, when calculating exposure towards a given counterparty or issuer, the calculation needs to take into ac-

16   GapGemini: 'Architecting Enterprise Data Management'.
17   Shane McGriff, GapGemini: 'Three Under-Appreciated Mandates for Enterprise Risk Management in the Next Decade'.

count all business with that company. This includes equity positions in the company's stock, positions in bonds issued by the company, over-the-counter (OTC) derivatives contracts with the company, derivatives with underlying assets issued by the company, cash and security collateral, unsettled trades where the company is counterparty, lending agreements with the company and so on. As can be seen it can be quite complicated to collect and consolidate all necessary information in order to be able to make the calculations. There are different reasons as to why it can be a significant challenge for some organisations to consolidate this data. It may be that different asset classes are managed and traded using different systems, some mandates, sub-strategies within a fund or portfolios may be managed by external portfolio managers and back office operations may be outsourced to one or more custodians or TPAs. Further, if all this information is stored in multiple systems, there will be an operational risk associated with collecting the information via interfacing the systems.

Information details level

Having ensured that all information is available in a consolidated view for risk calculations, it is also important to ensure that the data actually has the necessary granularity or detail level required for risk management. Requirements differ from department to department, so if data is consolidated across different systems used by different departments, they may not be at the same detail level. The details required to settle and match a trade are not the same as those required for portfolio accounting which is again different from the requirements of middle office functions such as risk and performance. For example, if a multi-manager fund has three external managers, the accounting department wants to see one position per security, whereas risk and performance needs to split positions per manager, in case more managers have invested in the same security.

The recent market environment with increased pressure on returns means that investment managers increasingly lend out securities, often via an agency providing that service. The increased focus on risk management has taken this trend further, so many investment managers now build their own lending business in order to reduce risk on the agency as well as to increase margins. At the same time collateral is increasingly used not only in lending agreements but in almost all OTC derivatives agreements. All this means, that the investment manager needs to keep track of his positions at a level of detail that was not necessary earlier. It must be possible to see if a position, or part of a position, is used as collateral.

Timely access to information

Given that information is available at the required level of detail and can be consolidated, the next challenge is to have access to the information when needed. Data may have to be collected from multiple sources, such as different internal systems or multiple custodians or TPAs. Online access to updated data is likely to be a key requirement as a consequence of the financial crisis. There have been several situations such as both the Lehman Brothers and Bear Sterns bankruptcy that took place over a weekend. In the case of Lehman, the International Swaps and Derivatives Association (ISDA) offered a special trading session on Sunday 14 September 2008 where market participants could offset positions in various derivatives against each other. In that case it was absolutely crucial to have access to correct weekend risk positions at short notice.

Flexibility

Another challenge is to keep enough flexibility to cope with changes. As discussed earlier, even though we, in the short term, will probably see an overweight in traditional assets traded on liquid markets and therefore relatively easy to value, the medium-to-longer term is likely to show continued investments in derivatives in the search for alpha. The derivatives may change, as seen in the recent increase in the demand for volatility and inflation products, but the requirements for transparency in risk management of them will be the same. The challenge is therefore to cope with the continuous introduction of new products and instruments that must be implemented swiftly to ensure short time-to-market.

Compliance control

Finally, as discussed earlier, a holistic approach to risk management integrates risk management and compliance. An enterprise risk management framework should consider risk assessment and compliance both on a pre- and post-trade basis. It must be ensured that portfolios are in compliance with the defined risk tolerances. Compliance rules are defined according to legislation, client restrictions or internal guidelines. Compliance should be monitored on several levels. Portfolio managers and traders must ensure on a pre-trade basis that trades do not violate any compliance rule. If a breach is overridden, it must be established who took the decision and why. At the same time, movements in the market may lead to breaches even without trading, so this must also be monitored on a regular basis. While many organisations have implemented pre-trade compliance validation, in some cases the solutions do not cover all relevant rules. For example, the US wash sales rule does not allow for selling and buying back the same security within 30 days. It is allowed, but it has consequences for the taxation of capital gains. Similar rules exist in other countries, for example to avoid dividend stripping, the purchase of a security just before dividend payout and sale immediately after payout. It is used for tax arbitrage because of different taxation of capital gains and dividends in different countries. Likewise, it is not usual to validate compliance to VaR tolerance levels on a pre-trade basis, not to say for Monte Carlo valuations. So even though pre-trade compliance is generally used among professional investment managers, it does often not cover all rules.

Reporting and disclosure

The driver for improved reporting and disclosure comes from the pressure from leg-islators, clients and the public to gain insight into investment managers' investment processes to ensure that they comply with legislation and client agreements.

Reporting infrastructure

As a minimum, investment managers must have a reporting infrastructure that is able to deliver reports as requested by their clients and regulators, and some may take the opportunity to invest in this area to gain competitive advantage.

Consolidation of information

The first challenge is similar to the one found under risk management that is, to be able to consolidate all data necessary for reporting. If data are stored in multiple sys-tems the reporting infrastructure must be able to access and consolidate necessary information.

Consistency in reporting

Tightly connected to this is the second challenge of ensuring consistency in reporting. If information must be collected from multiple systems, it is likely that each of these systems uses its own calculation models, parameter sets, aggregation methods and also they might not work on the same underlying data which requires careful reconciliation between sys-tems. For example, systems might use different methodologies for including unsettled trades and calculating key ratios based on different underlying security parameters. For non-additive key ratios like VaR or time weighted return (TWR) the aggregation levels must be synchronised between the systems to ensure consistent reporting which means that this is something that must be decided before the calculation, not after.

Documentation

The next challenge is new requirements for documentation. This could for example be documentation of best execution according to MiFID, which requires the asset manager to be able to document the client's execution price at exact trade time against competing bids in the market. The trend towards market fragmentation complicates this further, because it may be difficult to find the actual market price across all mar-kets. Another aspect is the need to document compliance to legislation and client restrictions at all times, which requires the investment manager to keep detailed com-pliance validation results and audit trails.

Reporting frequency

Another challenge for the investment manager is the expected increase in reporting frequency. Clients and regulators are likely to demand access to information with a much higher frequency than today. Currently it is not uncommon that some clients receive statements of their portfolios once a month or perhaps even once every quarter. The reporting process is often a time and resource consuming process based on heavy batch calculations, reconciliation, validations on multiple levels not to mention the distribution process. Many investment managers will find it challenging to increase the frequency of this reporting and a requirement for online reporting would be a complete change of operations for many.

Distribution channels
The final challenge has to do with distribution channels. Clients will ask for online access to reports via the Internet and regulators will also be requesting increasingly automatic reporting in order to enhance their automation of processes. While e-mail and file-based electronic distribution of reports is already widely used, web-based reporting may impose a challenge and change of operations for some.

Legislation and market standards
The challenges connected to addressing the business driver 'legislation' partially overlaps the challenges for the other business drivers. Legislators are expected to practice more scrutiny when assessing financial institutions both with regards to risk types that has traditionally been in focus and also to focus more on other risk types such liquidity risk and systemic risk. For example, it is expected that operational risk will receive greater attention again in light of recent fraud scandals such as that at Société Générale.[18]

This business driver is not only about legislation but also about market standards decided by industry associations and market practice groups such as the International Securities and Derivatives Association (ISDA) or Operations Management Practice Group (OMG). For example, from OMG there is pressure for higher derivatives matching and submission rates on electronic confirmations in order to address operational risk. While this initiative is not driven by legislation, investment managers will still have to comply with the general accepted market standard.

Transparency
Legislators will demand transparency in the reported figures. The challenges are the same as mentioned previously, for instance transparency in valuations, risk decomposition, risk calculations and market data used.

Timely delivery of information
Another of the expected challenges is shortened deadlines for responding to legislators. This demands that information can be consolidated and reported in the timeframes requested and distributed in the format demanded by the legislators.

---

18   PriceWaterhouseCoopers: 'Operational risk: an alternative challenge'. September 2008.

Fig. 4. The relationship between operating model, enterprise architecture and IT engagement model. Enterprise architectures should be defined based on the organisations' operating models.

Operating models

Operating models for an investment manager can be defined to different degrees of detail. In any case the starting point would typically be to define how the organisation does business, that is the overall operational value chain of the investment manager. This includes tasks and processes as well as the operating units involved. One particular aspect to take into account is that many investment managers have outsourced part of their operations. The most popular areas of outsourcing are back office functions such as fund accounting and tax reclaims among others. Other functions such as performance measurement and attribution, risk measurement and client reporting are typically considered closer to the core business of the investment manager and are less likely to be outsourced.[21] However, this may differ from investment manager to investment manager. Some offer asset allocation or fund-of-fund management advisory services and outsource portfolio management for specific asset classes or regions which are therefore not considered core competence. Likewise, many hedge funds outsource all operations including post-trade operations, risk and performance reporting, systems operations and so forth to their prime brokers. Yet again, other investment managers are part of large companies that offer a range of products and services.

This means that the operating units of the operating model can both be internal and external. There are various definitions of the basic value chain of investment managers.[22] In particular there are often different definitions of which tasks and processes belong to the middle office and which belong to the back office. Figure 5 shows a high level example of the main tasks in the value chain including illustration of various degrees of outsourcing.

Typical enterprise architectures

One of the most debated subjects within enterprise architectures and IT solutions is the question of best-of-breed versus integrated solutions. In short, the advantage of best-of-breed solutions is that they can provide richer functionality to specific areas of the operating model. On the other hand, it introduces challenges in terms of multiple systems and duplicate entry of the same data, interfaces to other systems via middle-

21   CityIQ and Northern Trust: Investment Administration Outsourcing Survey 2008.
22   James Wolstenholme, Gartner: 'From Investment Managers to Hedge funds: The operational landscape'.

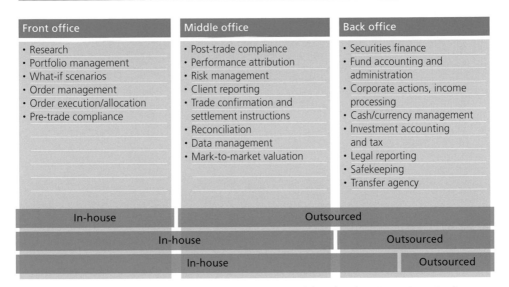

Front office	Middle office	Back office

• Research
• Portfolio management
• What-if scenarios
• Order management
• Order execution/allocation
• Pre-trade compliance
	• Post-trade compliance
	• Performance attribution
	• Risk management
	• Client reporting
	• Trade confirmation and
	  settlement instructions
	• Reconciliation
	• Data management
	• Mark-to-market valuation
		• Securities finance
		• Fund accounting and
		  administration
		• Corporate actions, income
		  processing
		• Cash/currency management
		• Investment accounting
		  and tax
		• Legal reporting
		• Safekeeping
		• Transfer agency

Fig. 5. High-level view of value chain for investment managers. The value chain is starting point for definition of operating models.

ware, redundant data storage and in general a more complex infrastructure. There is also need for specialised knowledge for every best-of-breed system, multiple vendor management and so on. A typical high-level example of a best-of-breed enterprise architecture is shown in figure 6. A number of trading systems and platforms, perhaps one per asset class, connect to custodians or TPAs for post-trade processing and fund accounting. The architecture contains a number of additional specialised systems for tasks such as risk management, performance, treasury and cash management, as well as miscellaneous spreadsheets, used by portfolio managers for portfolio optimisation or middle office for management reporting and so on. A number of data vendors supply security, reference and market data for valuations and the like to each application that needs the data. Systems are integrated by use of middleware.

If we compare the best-of-breed approach to some of the strategic challenges described earlier, we can make the following observations. In terms of ERM and consolidation of data there are restrictions to how this architecture can solve the challenges. There is not one repository where all position and transaction data can be consolidated across asset classes, managers, TPAs and others to give the full overview of exposure. Integration and duplication of data increase operational risk. Access to information in a timely manner depends on its integration. A reporting infrastructure would again have the same data consolidation issues and consistency in reporting would be a challenge. Data quality and EDM are not addressed in this architecture, therefore each business function would use its own data, calculation methods etc. Flexibility is dependent on how easily a new application can be integrated into the existing structures, mapping to the existing applications via the middleware.

The benefit on the other hand is the possibility to use best-of-breed applications for risk calculation, as for example access to exactly the risk models, complex valuation methodologies and decomposing of risk that the business requires.

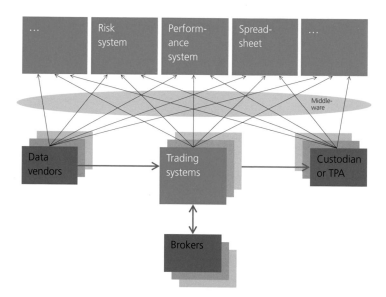

Fig. 6. Typical best-of-breed architecture

Integrated systems provide solutions to some of the shortcomings of best-of-breed architectures such as one data repository and no interfacing between internal systems. An integrated architecture is based on one common system for all business functions, where the priority is to have just one system, one technology, one vendor, one data repository and so forth. A high-level example of a fully integrated enterprise architecture is shown in figure 7 consisting of one system or platform for all business functions. This enterprise architecture would typically integrate to data vendors for security, reference and market data, to brokers for trade execution, and to the custodian for settlement and reconciliation.

If we compare the integrated approach to the strategic challenges, it shows that the issues with data consolidation and consistency are overcome since everything is based on the same application.[23] Information is often accessible either directly and online from the database or via batch calculations that typically can be run ad-hoc if needed.

On the other hand, the valuation models and risk calculations may not be as advanced as the best-of-breed applications, and may therefore not fulfil all the business requirements. Likewise, even though the underlying infrastructure for reporting may

---

23   In this context, it is worth noting that some integrated solutions actually consist of different applications offered by the same vendor or as a variant, originally different applications that are integrated by the vendor. In these cases, it is not necessarily guaranteed that the integrated solution uses one common data repository or that it uses the same underlying functions to do the same calculations in different parts of the application.

be in place[24] the reporting capabilities may not correspond to the best-of-breed systems. In general the individual business functions are typically not as functionally rich as comparable best-of-breed systems, so the investment manager must decide if the functionality is good enough so that the benefits outweigh the disadvantages.

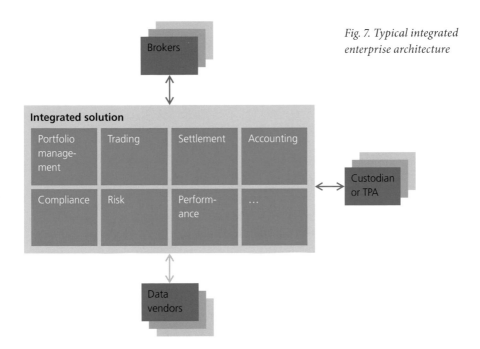

Fig. 7. Typical integrated enterprise architecture

The design of enterprise architectures is in many cases not a question of either or. For many larger investment managers, the enterprise architecture today consists of several systems that integrate in a combination of the two architectures described above. Even though this is the case, the task for investment managers going forward will be to define an enterprise architecture that has a stable core, which has the same basic structure over time, where specific functions can be integrated by observing the defined concepts and structure of the core. There seems to be a trend that more and more investment managers are turning their focus to this as a main strategic area and are adopting an integrated systems approach as the core of their enterprise architecture and then filling the gaps with best-of-breed systems.[22] Figure 8 shows a high-level picture as an example of a combination of best-of-breed and integrated enterprise architectures. The core of the architecture integrates to data vendors as well as to custodians or TPAs, the level of integration is decided by the operating model chosen by the investment manager. Ideally all data vendors should integrate directly into the core, which means that the core contains a full data set that is common to all business functions. It is, however, sometimes the case that a specific business function

---

24   This topic typically involves a fundamental question about reporting from an 'operational' database or a database designed for reporting, for example a data warehouse.

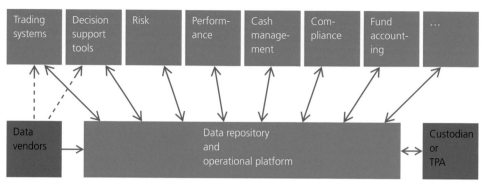

Fig. 8. Enterprise architecture with core and add-on best-of-breed systems

needs specific data that are unique for this function and is therefore not used in other business functions including reporting. In other cases, such as with real-time market data for example, permissioning issues prevents the distribution of data across several business functions and users. Therefore, there may be exceptions where data vendors feed directly into the business functions rather than into the core. However, in general each specific business function integrates to the core thereby using the common data. For this to be possible, the core of the enterprise architecture should contain basic or core data directly fed by data vendors but also, just as important, it should contain business logic for calculating derived data as needed by the business functions. For example, market values and other portfolio key ratios could be calculated by the core so the business functions all are based on the same set of key ratios calculated using the same parameters and business logic.

A comparison with the strategic challenges shows that this architecture potentially addresses many of the issues.

By having all data in one common repository, the consolidation issue is addressed. The information details issue is also addressed if the core is able to handle different levels of information needed by the different business functions at the same time. In terms of quantification of risk, the architecture allows for use of specialised risk applications, as long as it integrates into the core. Since both trading systems and data vendors are integrated directly to the core, the core contains updated positions and market data to be used by the business functions. The architecture facilitates maintaining high quality data, which however, requires good data management tools like golden copies, four-eye principles and other types of comparisons and validations. If defined thoroughly, the core allows for flexibility of adding new functions and financial instruments without changing the basic structures.

The reporting infrastructure also benefits from this architecture because all data are consolidated in one place so the same key ratios are not calculated in different applications using different calculation models and parameters. The architecture allows for flexible and advanced reporting using specialised solutions for example for web-based reporting. The model also addresses flexibility in terms of operating models. For example, if the current operating model is to outsource fund accounting to a TPA, this enterprise architecture allows that this could at a later stage be in-sourced using an in-house function or application, as long as it is integrated to the core.

As shown, this enterprise architecture potentially solves the issue with either a pure best-of-breed or integrated approach. However, the implementation of the core can be done in several ways.

The first option is to outsource the core data master to the custodian or TPA. If the investment manager uses just one global custodian or TPA it can be argued that the TPA already keeps track of all the investment managers positions, so why not let the TPA be the central repository and feed specialised systems from there? This may be a solution for some, however, the investment manager needs to consider a number of issues. Does the TPA allow for access to data in a timely manner? The financial crisis has shown that prudent risk management requires online access to updated and validated position data, sometimes also during weekends. If the investment manager uses more than one custodian or TPA, and this is often the case for discretionary mandates or managed accounts, there still needs to be a consolidation across TPAs which goes against the basic idea. Another aspect to consider is, if the TPA can provide data at the level of detail and quality required by the different functions. The level of detail for back office processing may not be sufficient for front or middle office analysis and requirements may vary over time. Generally, the investment manager needs to carefully consider the fact that whereas they themselves typically want flexibility to scale their business by differentiating their services from their competitors, TPAs scale their business by standardisation. If the investment manager wants to take advantage of cost efficient operations by the TPA, they should be aware of these underlying forces pulling in opposite directions. Therefore, investment managers who have outsourced part of their operations to a TPA, should consider having an enterprise architecture with a core data master.

The second implementation option is to use the legacy accounting system as core. While the accounting system may consolidate information about all asset classes, the main drawback is that the requirements of the accounting department and the requirements of the front and middle office are very different. An accounting system is account-based and the level of detail for these accounts is determined by the required accounting standards which may not be the same as for other departments. For example, for a life insurance company there may be one profit and loss (P&L) account for bonds as required by the accounting department, whereas front/middle office needs to decompose P&L into issuers, regions and so forth. Generally, accounting systems organise data according to accounts whereas front and middle office wants to view data according to positions. So, in general, accounting systems are not suitable as a core data master.

The third implementation option is to build a data warehouse and use that as the core. This is a widely used approach at many investment managers. One of the issues with this solution is that it may be a difficult task to build a data warehouse/data model that fulfils the requirements from all departments. As described earlier, the requirements from front, middle and back office differ substantially and the data model needs to take everything into account. Another challenge is that data warehouses are fed with data from operational systems via an extract, transform, and load (ETL) process. Because of this there is always some kind of latency in the data of the data warehouse.

Conceptually, a data warehouse is for data storage only. In many cases core data needs to be enriched or further processed as derived data to be used by receiving systems. For example, a performance system typically needs daily market values and transactions/movements. If the TPAs feed the core with positions and data vendors feeds prices and FX rates, there must be some logic for calculating market values. One way could be to use some prioritisation logic that determines whether to use one price type or another from one vendor or from another, such as using bid prices for long positions but only if there is a bid from today, otherwise use close price not older than two days etc. In case of OTC derivatives the situation is even more complex because there is no public market price, so the market value must be calculated based on theoretical pricing. It may also just be that the investment manager wants to use different FX rates than those provided by the TPA, or that some clients prefer one price source while others prefer another, for instance according to their benchmark.

In summary, this suggest that the core should contain some business logic for valuation and other calculations. By this approach, data used across different business functions are calculated in common consistent ways in the core instead of the individual systems. The valuation models used by performance and risk for client reporting should be the same. Individual systems will still contain calculation logic that is very specific for that system, such as TWR for performance or VaR for risk measurement. For clients operating on a global scale, there are other issues to consider. Does the data warehouse provide functionality for 24x7 support across multiple time zones? Issues like global/local timestamps, local end-of-day are likely to arise. It is also important that operational procedures are supported by the data warehouse. This includes things such as permissioning, audit trail, monitoring of servers and processes among other things. The final aspect to consider is perhaps the most important. Requirements from the business may change over time and even if designed correctly this requires updates and maintenance to the data warehouse, which can be costly. There is a certain degree of overlap between data warehouses and operational systems and duplicate functionality may be developed. In general, the operating systems should aim to deliver functionality and data warehouses only store data and do not further process data.

The fourth option is to use an integrated investment management system as a solution for the core. Integrated investment management systems are typically so-called 'operational systems' meaning that the data model is optimised for operational use by the business functions which are covered by the integrated solution. The benefits of this approach are that, within the scope of the integrated system, the core contains all necessary data. Further, integrated systems typically contain business logic in contrast to other solutions such as data warehouses. This means that the core contains all core data such as security information, reference data and market data, often on a very detailed level including time series for historical changes, security information as basis for theoretical pricing or cash forecasts, detailed classifications and other information. This is simply because this information is needed for operation of the business functions covered by the integrated system. This means that the integrated system may contain a data model for core data as well as for derived data including business logic, ensuring that derived data are calculated by using the same functions and core data across business functions. Some integrated systems also contain support for advanced data structures, which means that business functions can be more tightly integrated into the core. For example, the core data master may provide core and derived data in the form of positions, market values and transactions for business functions like risk and performance. The result of the risk and performance calculations and analytics may then be returned and stored in the core in pre-defined data structures, for example with respect to the aggregation levels, which is especially important for non-additive key ratios, to be used by the reporting function. Further, integrated investment management systems often support a range of operational functions such as permissioning and audit trail as well as system operations tools such as monitoring of servers and batch jobs. The most widely known issue with this implementation approach is that integrated systems databases often are primarily designed for operations and not for reporting. This may imply that a reporting database is needed alongside, or rather as an add-on to the core. Another challenge with using an integrated system as a core is the flexibility and ease of integrating best-of-breed solutions into the core. This requires the integrated system to be 'open' and contain excellent integration tools.

Summary

We have identified three current main business drivers for investment managers: risk management, reporting and disclosure as well as compliance to legislation.

The main strategic challenges for investment managers in relation to the business drivers were described from an IT perspective. Some of the main challenges are the ability to consolidate data to get a full overview of risk, to ensure that there is enough detail for analysing risk properly, to ensure online access to positions, to ensure the quality and consistency of the underlying data and calculations and to ensure transparency in valuation and calculation process.

We discussed that an IT enterprise architecture should not be designed with IT as its starting point but rather based upon the business' operating model. The key for investment managers should be to define an enterprise architecture with a stable core, with main data structures and workflows that remain unchanged when adding new functions or financial instruments on to it. A number of typical enterprise architectures were analysed and compared as to how they met the strategic challenges.

The analysis showed that no matter which operating model investment managers have, they should seek to have a core data master as part of their enterprise architecture to meet their strategic challenges. The core ensures that all information can be consolidated, is available online and that underlying data and calculations are consistent. The core should not only be a data repository, but should also contain business logic for calculating derived information and it should also contain functionality like permissioning and audit trail to make it an operational system.

A number of different ways to implement an enterprise architecture with a core data master were discussed. The different approaches to defining a core all have strengths and weaknesses. The most optimal approach is suggested to be a solution where the core is based on an integrated solution as an operating system that allows for integration of best-of-breed where appropriate. This core may be complemented by an embedded reporting data model to overcome the challenges of reporting from a database designed for operational use.

References

Aden: 'Enterprise Architecture Demystified', *Government Technology,* 24 September 2008 (http://www.govtech.com/gt/articles/418008?id=418008&full=1&story_pg=1).

Balin: 'Basel I, Basel II, and Emerging Markets: A Nontechnical Analysis', The Johns Hopkins University School of Advanced International Studies (SAIS), Washington DC 20036, USA. May 2008.

Bruel, Stephen, TowerGroup: 'Understanding Securities Industry Risk by Using the TowerGroup Risk Management Matrix'. March 2009.

CityIQ and Northern Trust: 'Investment Administration Outsourcing Survey 2008'.

Ernst & Young: 'Managing Information Technology Risk. A global survey for the financial services industry' (2008).

GapGemini: 'Architecting Enterprise Data Management 2007'.

Gartner: 'Key Issues and Research Agenda for Banking and Investment Services 2009.'

McGriff, Shane. GapGemini: 'Three Under-Appreciated Mandates for Enterprise Risk Management in the Next Decade', 2009.

Needleman, Peter. Towers Perrin: 'The Last Word: The Financial Crisis: Is mark-to-market to blame?' 2008.

Press release by US Federal Reserve 2009 (http://www.federalreserve.gov/newsevents/press/bcreg/20090507a.htm).

PriceWaterhouseCoopers: 'Operational risk: an alternative challenge'. September 2008.

Ross, Weill, Robertson: 'Enterprise Architecture as a strategy. Creating a Foundation for Business Execution', 2006.

Schröter, Marc, SimCorp: 'Outsourcing and Risk Management, time to redefine requirements?' *Journal of Applied IT and Investment Management,* vol 1, no 1, 2009.

Schröter, Marc, SimCorp: 'White paper on Extended Enterprise Data Management', 2009.

Shahrawat, Dushyant and Dayle Scher, TowerGroup: '2009 Top 10 Business Drivers, Strategic Responses, and IT Initiatives in Investment Management'.

SimCorp: 'Global Investment Management Risk Survey 2009'.

The 2009 Ernst & Young business risk report. The top 10 risks for global business.

The Compliance Consortium: 'Governance, Risk Management and Compliance: An Operational Approach'. May 2005.

Wall Street & Technology Blog 14 October 2008: 'One Way to De-Mystify Derivatives Valuation'.

Willsher, Richard: 'Credit ratings: IT in the spotlight', *Journal of Applied IT and Investment Management,* April 2009.

Wolstenholme, James, Gartner: 'From Investment Managers to Hedge funds: The operational landscape'. 2008.

Regulation and governance

OVERVIEW

6 Corporate governance and the financial crisis

by Steen Thomsen

Professor Steen Thomsen, Ph.D., is director of the Center for Corporate Governance at Copenhagen Business School. He specialises in corporate governance as a teacher, researcher, consultant, commentator and practitioner. His main publications include more than 25 international journal articles and three books. He has served as a board member of several companies and is currently a non-executive chairman of two consulting firms. Contact: st.int@cbs.dk

The financial crisis 2007–2009 is often partly attributed to bad corporate governance. For example, the OECD Steering Committee on Corporate Governance argues that "the financial crisis can to an important extent be attributed to failures and weaknesses in corporate governance" (Kirkpatrick, 2009). The standard story appears to be that weak boards tolerated the rise of a culture of greed and excessive pay, which led financial executives to take the risks that ultimately caused the financial crisis (see for instance Obama 2009). In this paper, I examine the rationale for this assertion.

The literature is still quite limited. Cheffins (2009) and Adams (2009) provide some of the first academic research on the subject. Cheffins examines governance characteristics of 37 firms removed from the S&P 500 index during 2008, including companies like Lehman Brothers, Countrywide and Bear Stearns. He finds little evidence of Enron-style fraud, but notes that boards were criticised for lacking sufficient financial expertise and for not challenging the CEO's dominant position. Moreover, pay in many cases was very high and criticised for being excessive. However, boards did react when financial distress became apparent and in many cases replaced top executives. Thus, he concludes that corporate governance worked reasonably well and warns against fundamental reforms on this shaky foundation.

Adams (2009) examines governance structures at organisations ranging from banks and non-bank financial institutions to non-financial firms. She finds that bank boards are not obviously worse than those of non-financial firms. In fact their governance appears to be as good or better in terms of board independence and other important characteristics. Moreover, banks with more independent boards were more, not less likely to become financially distressed during the crisis.

In this paper, I employ a different approach by relating current governance research to the issues raised by the financial crisis. Corporate governance can be defined as the mechanisms by which companies are directed and controlled. These mechanisms include active ownership, board supervision, managerial incentives, company law and informal mechanisms such as reputation. The question then is whether and to what extent the financial crisis was caused by failure in one or more of these mechanisms. This paper reviews what is known about the performance of these mechanisms during the crisis. I focus mainly, but not exclusively, on banks and financial institutions.

Agency problems in financial institutions

The classic agency problem, which corporate governance is expected to address, is that between a group of weak, dispersed shareholders and a strong CEO. The shareholders want maximum returns for a given level of risk, while the CEO has an interest in high pay, corporate expansion, high levels of expenditure and absence of control (Williamson, 1964). Corporate governance mechanisms then work to control CEO behaviour (Tirole, 2006). For example, boards can negotiate sufficiently low executive pay and fire a non-performing CEO. Shareholders can rebel at the annual meeting and elect a new board or a hostile raider such as a private equity fund can take control, fire the CEO and clean up the company. Other agency problems like fraud, self-dealing or insider trading are assumed to be handled by law or regulation. Yet others are supposed to be overcome by performance-related pay. Even reputation can be a governance mechanism because it induces managers to perform well for fear of reputational loss. However, despite many governance mechanisms agency problems cannot be eliminated altogether (Tirole, 2006).

In principle, agency problems in financial institutions are no different from those of other corporations. They are known to depend on the nature and level of information asymmetries between managers and shareholders, the level of uncertainty in the business as well as the risk preferences and incentive-sensitivity of managers. However, financial institutions are special because they have many residual risk takers, so-called 'principals', including shareholders, debt holders, who are particularly involved in financial institutions because of their high degree of leverage, and governments, which pay the bills in bailout situations, not to speak of the general public who suffer during system breakdowns. The systemic character of financial institutions has led to extensive regulation of their governance, including, depending on their jurisdiction, monitoring by government agencies, mandatory procedures for board approval of major loans, government appointed board members, limitations on directorships in other corporations and solvency rules.

However, during the past decades agency problems in financial institutions may have become worse. Their increasing size and complexity has made it hard for boards, shareholders and regulators to understand what they do. This refers to such things as the creation of financial supermarkets and their increasing globalisation as well as the increasing sophistication of financial products like credit derivatives, swaps and so forth (Acharya et al. 2009). Complexity has grown because deregulation made it possible for banks and other institutions to diversify into related activities like mortgages or insurance and to organise a substantial share of their activities in off-balance sheet operations. Moreover, financial globalisation has made it possible to diversify risk to a greater extent, which has added to the complexity while reducing the incentives to monitor individual institutions and products. Under Basel II, banks were given the responsibility for assessing the risks of their loan portfolios. The assessments were based on statistical models which were subject to approval by regulators. Self-reporting however involves obvious agency problems because managers could have a short-term incentive to free up capital and the regulators found it difficult to assess their increasingly sophisticated risk models, which were anyhow based on fairly recent data from the boom period. The problems were aggravated because mark-to-market accounting under the new international accounting standards forced many of them to reduce their hidden reserves and post better short-term results. All of this made financial institutions increasingly difficult to control and shifted much responsibility to boards and other governance mechanisms.

Similar, although perhaps less severe, problems arose in non-financial firms although the risks they took were related to overinvestment and borrowing rather than to lending and guaranteeing complex financial products. High economic growth, record profits and abundant credit made it easier to finance CEOs' desire for expansion and bid up share prices on the tacit assumptions that growth would continue, short-term funding would always be available and share prices would continue to rise. New financial institutions like private equity and hedge funds contributed to the trend. This put a strain on corporate governance to restrain executive growth ambitions and limit unsound leverage.

Executive pay

Executive pay is widely blamed for contributing to the crisis. This section will examine the justification for this assertion. Quoting Barack Obama (2009): "We're going to examine the ways in which the means and manner of executive compensation have contributed to a reckless culture and quarter-by-quarter mentality that in turn have wrought havoc in our financial system".

The standard argument is that excessive risk taking in banks and other corporations was partly caused by dysfunctional managerial incentives such as stock options and short-term bonuses. This is a view shared by the OECD (Kirkpatrick, 2009) and the Financial Stability Forum (2008, 2009). For example, the Financial Stability Forum (2008) states that "compensation schemes in financial institutions encouraged disproportionate risk-taking with insufficient regard to longer-term risks". This argument has led to a series of proposed policy changes. For example that "Compensation must be adjusted for all types of risk … Two employees who generate the same short-run profit but take different amounts of risk on behalf of their firm should not be treated the same by the compensation system … Compensation outcomes must be symmetric with risk outcomes … Bonuses should diminish or disappear in the event of poor firm, divisional or business unit performance … Payments should not be finalised over short periods where risks are realised over long periods" (Financial Stability Forum, 2009). However, the Financial Stability Forum declines to take positions on the level and performance orientation of executive pay arguing that "one size does not fit all".

The case for intervention has been argued by Bebchuck and Fried (2004), who see high pay levels in the US as evidence of rent extraction, that is that managers have been able to extract excessive pay substantially above their marginal productivity. The high compensation levels of failed financial institutions have been cited as anecdotal evidence (Chung et al. 2008). For example Angelo Mozilo, CEO of Countrywide, reportedly received $52m in 2006, branding him the 'poster child of financial services excess'. In 2007 he got only $10.8m, but made a further $121.5m from cashing in stock options. Stan O'Neal, former CEO, Merrill Lynch, got $162m before the company was sold to Bank of America. Richard Fuld of Lehman Brothers got $34.4m in 2007 before his company went bankrupt. Charles Prince of Citigroup is said to have received $25.9m in 2006, plus severance pay of $40m in 2007. However, others have argued that the 'fat cat' theory is too simplistic (Holmstrom 2005, Hubbard 2005, Murphy, K., & Zábojník, J. 2004). For example, CEO pay appears to have increased across the world reflecting a scarcity of talent, and the demand for talent in financial services during the boom years was no doubt exceptional. Moreover, the perceived value of additional risky equity compensation may have been quite low given their already substantial wealth and effort levels.

Moreover, the extremes are obviously extremes. Taking into consideration that banks are large firms, they should be compared with other large firms Adams (2009) finds that bank CEOs have lower overall pay and less equity-based pay than their counterparts in non financial firms. Thus it is not clear that bank managers' pay is particularly out of line. However, she did find that equity-based compensation was associated with an increased probability of government bailout.

In any case, the responsibility for executive pay rests with boards. Jack Welch has argued that pay excesses are caused by weak boards (Welch and Welch, 2006). If executive pay has been excessive, attention is therefore directed at board structure.

Board supervision

Boards of banks and other corporations could in principle have prevented the excessive risk taking which led to the credit crunch. According to the OECD, risk policy which is 'a clear duty of the board', failed because of insufficient information and insufficient monitoring by bank boards (Kirkpatrick, 2009).

Unfortunately, there is no consensus on what constitutes a good board. Hermalin and Weisbach (2003) review empirical studies of board structure and company performance and find no systematic effects except a negative association between board size and firm value. Even this negative finding has subsequently been questioned by Coles et al. (2007) and particularly for banks by Adams and Mehran (2007). Staggered boards and busy boards with many interlocks to other boards have also been claimed to have negative effects, but again there are significant problems involved in assessing the direction and magnitude of causal effects. One obvious problem is endogeneity: board structure is influenced by many other factors, which may also co-vary with company performance (Adams et al. 2009). Another, highly pertinent problem is that it is difficult to monitor risk when measuring bank and company performance. Many of the high performers of the bubble period turned out to be financially fragile because they had taken on too much risk. Thus the financial crisis has made it painfully clear that accounting numbers and market values were often not trustworthy because they did not take into consideration substantial underlying risks.

Adams (2009) finds that bank governance is generally fairly good compared to non-financial firms. She finds that banks have on average more independent and larger boards, fewer outside directorships, higher total CEO compensation and lower director compensation than non-financial firms, although these results turn out to be sensitive to firm size. Allowing for size, non-bank financial firms have less independent and smaller boards, fewer outside directorships and lower director compensation than non-financial firms. Altogether, she concludes that the governance of banks and financial institutions was not obviously worse than that of other businesses. This is perhaps not surprising given the intensive regulation and public visibility of financial firms. See for example the fit and proper tests required by board members in financial institutions, FSA 2004. It is not clear, therefore, that the financial crisis was caused by particularly bad governance of financial institutions. It may be that financial institutions were particularly exposed during the crisis or that higher standards of corporate governance are necessary in financial institutions precisely because they are occasionally subjected to financial crises. However, Adams (2009) also shows that the boards of banks receiving bailout money were more, rather than less independent than those of other banks. Thus independence is no guarantee of effective risk management. On the contrary, effective risk governance may require that non-executive directors know more about the financial industry, which makes it attractive to recruit former financial executives. The OECD argues that non-executive board members may have had insufficient expertise in banking and finance (Kirkpatrick, 2009). Adams (2009) found that bank non-executive director pay is some 30% lower than in non-financial firms, and Adams speculates that this may have made it difficult to attract highly qualified board members.

Altogether, it is not clear that the board structures of banks and other financial institutions are particularly bad as compared to other firms, but it is clear that their boards were not good enough to protect them from financial distress. In this respect they were a failure.

Information systems

Case studies of leading financial institutions demonstrate that the boards were often inadequately informed about their overall risk profile (Kirkpatrick, 2009). Moreover, risk management practices and executive compensation were often unrelated to and inconsistent with fundamental strategic decisions. One reason for this was no doubt the complexity of evaluating complex financial assets and liabilities. Another reason appears to have been the 'silo' approach according to which individual divisions of complex financial institutions managed their own financial risk without regard to the overall financial position of the institution. In addition, risks related to liquidity, reputation and system effects appear to have been underestimated. Thus, there is room for improvement in the information provided to company boards.

Nordgard (2009) reviews how information technology can strengthen emerging practices in governance, risk and compliance (GRC). He envisions that risk measures will be introduced at all levels in the organisation and quantified as 'risk budgets', that is to say, maximum acceptable risk. Compliance controls must check that different internal units stay within their risk budget. Reporting and suitable analysis will then allow efficient aggregation of information to assess overall risks at the board level, as well as efficient disaggregation of risk targets to individual units. However, he emphasises that the right foundations with regard to, for example, timely and accurate reporting must be in place before sophisticated analytical tools can create value.

Active ownership

Active ownership by self-interested shareholders might have prevented some of the value destruction during the crisis. Compared to non-financial firms, banks have quite dispersed ownership and therefore weak shareholders, partly because many agency concerns are already addressed by financial regulation (Demsetz and Lehn, 1985). Some have argued that empowering shareholders might provide better discipline in the future. For example it has been recommended that the US adopts a more

British governance model with greater accountability to shareholders. However, British banks have not done better than US banks during the crisis, nor is it clear that banks with higher ownership concentration have been better at handling risk. It is not evident that empowered shareholders would necessarily induce banks to take less risk, since there may be conflicts of interests between shareholders and debt holders. Shareholders may potentially gain by taking on more risk while debt holders lose.

It is interesting that some of the most highly leveraged institutions, the US investment banks, were historically closely held partnerships, in which high leverage was combined with a substantial equity stake by senior employees in the future of the business and thus counterbalanced excessive risk taking. In many institutions this structure was replaced by public listing and separation of ownership and management, which may have lessened the incentive for managers to take downside risk into consideration. Moreover, it may be no accident that many financial institutions across the world are not publicly listed companies, but cooperatives, financial mutuals and the like. While these firms may be less profitable during boom periods, they also appear to be more financially stable (Hesse and Cihák, 2007).

While past ownership structure may have been problematic, nationalisation of banks, which many countries have resorted to as a response to the crisis, is also far from ideal. Government ownership of banks is believed to be associated with slower subsequent financial development and lower growth of productivity and per capita income (La Porta et al. 2002). One important problem is that government ownership is often associated with political interference in lending decisions (Spanienza, 2004).

Capital structure

With equity to asset ratios of 10% or less, banks and financial institutions are highly leveraged compared to other companies. They are therefore *prima facie* cases of debt-governed corporations as envisioned by Jensen and Meckling (1976). Creditors have an obvious incentive to safeguard their interests by assessing and monitoring management in the companies that they finance, including, in particular, their risk policies. However, several factors may have diminished the vigilance of creditor monitoring. First, deposit insurance and the likelihood of government bailout, the so-called 'Greenspan put' may have acted as an implicit guarantee which dulled the incentives of creditors to monitor financial institutions and their management. Secondly, the 'originate to distribute' business model which involved repackaging, securitisation and reselling of loans to take them off balance sheets, may also have led to less monitoring. It is a paradox, however, that the risks of these loans turned out as matter of fact not to have been outsourced partly because put options and reputation risks brought them back to bank balance sheets when the crisis materialised. In fact a large part of the outsourcing went to other, similar financial institutions rather than to pension funds and other agents that might have been better able to carry the risk. Third, the increasing size and complexity of financial institutions and their products also made it more difficult for creditors to understand and monitor the risks they were getting into. Finally, outsourcing of the monitoring function to rating agencies appears to have failed.

The risks of high leverage in financial institutions are obviously exacerbated if they themselves invest in leveraged assets like hedge funds or commercial property. An added concern is that fair value, mark-to-market accounting and individual risk weights estimated over relatively short periods of time may provide financial managers and their boards with a false sense of solvency security and thereby increase their risk appetite.

Reputation

Reputation concerns can work as a corporate governance mechanism by deterring excessive compensation and risk taking or poor board monitoring (Adams et al. 2009, Fama 1980). It is well known that reputation can help overcome incentive problems in long-term relationships, but also that reputation is no panacea. Reputation is obviously not a very precise instrument when there are large information asymmetries and the media distort the flow of information. Holmstrom (1999) argued that reputation concerns could cause managers to become overly risk adverse. Walter (2009, this volume) tries to make precise the concept of reputation risk for financial institutions. He notes that financial institutions are particularly vulnerable to reputation risk because they depend so much on confidence and provides a specific illustration of the magnitude of operational risk.

It is unclear to what extent the crisis can be attributed to a dilution of reputation concerns. It is known, however, that reputation mechanisms may break down when information signals are distorted, when there are strong incentives for opportunistic behaviour and when players have short time horizons. It seems possible that the increasingly opaque nature of complex financial products has made it more difficult for shareholders and depositors to assess the risk profiles of financial institutions and to punish offenders with a decline in reputation. Moreover, it seems plausible that the short-term bias of pay structures in investment banking and other financial firms came to dominate the concern for long-term reputation.

Conclusion

Many of the popular stories of the financial crisis make little sense. For example, it is not very credible that the crisis was caused by a sudden epidemic of greed in the financial community. The major piece of evidence for this story appears to be that executive pay in some of the leading firms rose to very high levels, but this is insufficient. The alternative story that greed evolved over many years only to be exposed by the crisis indicates that it was caused by other factors such as easy money, deregulation and inadequate corporate governance. A simple alternative interpretation is that decades of rising asset prices led to expectations of continuing high returns, search for yield and increasing risk taking which was reinforced by performance-based pay.

Corporate boards failed to prevent excessive leverage both among borrowers and lenders. It is not evident that the corporate governance of financial institutions was worse than that of non-financial firms or that it suddenly changed for the worse before the crisis. However, financial institutions have doubtlessly become more difficult to control and agency problems have become worse because of their increasing size, complexity and leverage as well as financial deregulation. Moreover, the credit expansion following lax monetary policy in the period 2003-2006 and the global savings glut, no doubt reinforced moral hazard and adverse selection problems by pushing up asset prices and providing easy money for all sorts of projects. Evidently, the governance of banks and other financial institutions was not up to the challenge. Reforms and regulations are therefore to be expected.

Among the current proposals for reform we highlight the following:

1 Executive pay in banks and other financial institutions should be restructured to include more long-term components including restricted stock to be held over a long-time period, inclusion of debt as well as equity, bonus accounts which can be reduced if long-term performance fails to meet expectations and by monitoring risk in calculating capital costs when assessing managerial performance. Thus reform proposals do not aim to cap executive pay nor even its performance orientation but rather its time horizon.

2 Changes in ownership structure are in order. Government ownership of banks is known to be problematic. Nationalised financial institutions should therefore be privatised as soon as the macroeconomic climate allows. In the long term new ownership structures may be called for in order to counterbalance risk taking. For example, greater ownership concentration and more downside risk may to some extent counterbalance risk taking. Alternatively unlisted financial mutuals may be more financially stable than listed entities (Hesse and Cihák, 2007).

3 The boards of financial institutions should be upgraded to include more members with financial industry expertise, such as former bankers, as well as greater financial literacy. For example, if boards are to take charge of risk management or executive pay, new competencies are needed (Financial Stability Forum 2009).

4 Information systems should be upgraded to provide an overall view of financial and strategic risks that can be integrated into both risk management and executive compensation schemes. To govern, boards must have a comprehensive, accurate and

updated view of activities and risks including off-balance sheet activities and the like. If this is too complex to do in practice, financial institutions may be become ungovernable and the complexity should probably be reduced by restructuring.

5   Regulators should devise ways to provide incentives for creditor monitoring with the necessity of bailing out financial institutions in times of crisis. Increased solvency is not necessarily the solution because this may push retail banks to take on more risks to get a competitive return on equity.

With these reform proposals there is an obvious danger that regulators will attempt to fight the last war. The situation now is fundamentally different from what it was before the crisis. A generation of financial executives has learned some hard lessons, and there is currently little need to reinforce the lessons learned. For example, risk taking and leverage are not the problems that they used to be and curbing them further may do more harm than good to the rest of the economy. However, history shows that it is necessary to act now, while the memory of the crisis is still fresh, if major reforms in financial regulation are to take place.

References

Acharya, V. and M. Richardson (Eds.). Restoring Financial Stability: How to Repair a Failed System, Wiley, March 2009.

Acharya, V, Jennifer Carpenter, Xavier Gabaix, Kose John, Matthew Richardson, Marti Subrahmanyam, Rangarajan Sundaram and Eitan Zemel. 2009. Corporate Governance in the Modern Financial Sector. Chapter 7. Acharya, V. and M. Richardson (Eds.). Restoring Financial Stability: How to Repair a Failed System, Wiley, March 2009.

Adams, Renée. 2009. Governance and the Financial Crisis. Steen Thomsen, Caspar Rose and Ole Risager (eds.): Understanding the financial crisis: investment, risk and governance. SimCorp StrategyLab, 2009.

Adams, Renée B., Hermalin, Benjamin E. and Weisbach, Michael S. 2009. The Role of Boards of Directors in Corporate Governance: A Conceptual Framework & Survey. Charles A. Dice Center Working Paper No. 2008-21; ECGI – Finance Working Paper No. 228/2009; Fisher College of Business Working Paper No. 2008-03-020. Available at SSRN: http://ssrn.com/abstract=1299212.

Adams, R. and H. Mehran. 2008. 'Corporate Performance, Board Structure and its Determinants in the Banking Industry'. University of Queensland Working Paper, 2008.

Akerlof, George A and Robert J. Shiller (2009) Animal Spirits: How Human Psychology Drives the Economy, and Why It Matters for Global Capitalism. Princeton University Press.

Bebchuk, L. A. and Fried, J. M. (2004) Pay Without Performance: The Unfulfilled Promise of Executive Compensation, Harvard University Press, Cambridge, MA.

Bicksler, James L., 2008. The subprime mortgage debacle and its linkages to corporate governance. International Journal of Disclosure and Governance (2008) 5, 295–300.

Cheffins, Brian R., 2009. Did Corporate Governance 'Fail' During the 2008 Stock Market Meltdown? The Case of the S&P 500 (May 2009). Available at SSRN: http://ssrn.com/abstract=1396126.

Clementi, Gian Luca, Thomas F. Cooley, Mathew Richardson and Ingo Walter. Rethinking Compensation in Financial Firms. Chapter 8. in Acharya, V. and M. Richardson (Eds.). Restoring Financial Stability: How to Repair a Failed System, Wiley, March 2009.

Hesse, Heiko and Cihák, Martin. 2007. Cooperative Banks and Financial Stability(January 2007). IMF Working Paper No. 07/02. Available at SSRN: http://ssrn.com/abstract=956767.

Coles, Jeffrey L., Michael L. Lemmon, and J. Felix Meschke. 2007. 'Structural Models and Endogeneity in Corporate Finance: The Link Between Managerial Ownership and Corporate Finance,' 2007. Working Paper, Arizona State University.

Demsetz, H. and Lehn, K. (1985): The structure of corporate ownership: causes and consequences. Journal of Political Economy, 93(6): 1155–1177.

Fama, Eugene F., 'Agency Problems and the Theory of the Firm,' Journal of Political Economy, 1980, 88 (2), 288–307.

Financial Stability Forum (2008), Report of the Financial Stability Forum on Enhancing Market and Institutional Resilience (399 KB PDF), (Basel, Switzerland: FSF, 7 April).

Financial Stability Forum (2009), FSF Principles for Sound Compensation Practices (87 KB PDF). (Basel, Switzerland: FSF, 2 April).

FSA (2004). The Fit and Proper test for Approved Persons. Chapter 2. Main assessment criteria http://www.fsa.gov.uk/pubs/hb-releases/rel27/rel27fit.pdf. FSA Handbook.

Caprio, G., Laeven, L., and Levine, R. (2003): Governance and bank valuation, NBER Working Paper 10158.

Joanna Chung, Greg Farrell, Francesco Guererra and Saskia Scholtes. 2008. The Fallen Giants of Finance. Financial Times. 23 December 2008.

Hermalin, Benjamin E and Michael S. Weisbach(2003). 'Boards of Directors as an Endogenously Determined Institution: A Survey of the Economic Literature,' Economic Policy Review, April 2003, 9 (1), 7–26.

Holmstrom, B. (2009) 'Managerial Incentive Problems – A Dynamic Perspective,' Review of Economic Studies, January 1999, 66 (226), 169–182.

Holmstrom, B. (2005). Pay without Performance and the Managerial Power Hypothesis: A Comment. *Journal of Corporation Law,* 30(4), 703–715.

Hubbard, R. (2005). Pay without Performance: A Market Equilibrium Critique. *Journal of Corporation Law,* 30(4), 717–720.

Jensen, M. and W. Meckling, 1976, 'Theory of the Firm: Managerial Behavior, Agency Costs and Ownership Structure,' *Journal of Financial Economics 3,* pp. 305–60.

Kirkpatrick, Grant: The Corporate Governance Lessons from the Financial Crisis. *Financial Market Trends.* OECD 2009.

Murphy, K., & Zábojník, J. (2004). CEO Pay and Appointments: A Market-Based Explanation for Recent Trends. *American Economic Review,* 94(2), 192-196.

Nordgard, Kjell Johan. 2009. Governance, risk and compliance in the financial sector: an IT perspective in light of the recent financial crisis. Steen Thomsen, Caspar Rose and Ole Risager (eds.): Understanding the financial crisis: investment, risk and governance. SimCorp StrategyLab, 2009.

Obama, Barack, Speech on Executive Compensation. http://www.cnbc.com/id/29014485/.

CNBC.com 4 Feb 2009

Rafael La Porta & Florencio Lopez-De-Silanes & Andrei Shleifer, 2002. Government Ownership of Banks. *Journal of Finance.* 57(1), pages 265–301.

Sapienza, Paola. 2004. The Effects of Government Ownership on Bank Lending [Abstract], *Journal of Financial Economics,* May, 72(2): 357–384.

Thomsen, Steen. 2008. An Introduction to Corporate Governance. Djoef Publishing. Copenhagen 2008.

Tirole, Jean. 2006. Corporate Governance. Chapter 1. The Theory of Corporate Finance. Princeton University Press. Princeton.

Walter, Ingo. 2009. 'Reputational Risk and The Financial Crisis' in Thomsen Steen, Caspar Rose, and Ole Risager (eds): Understanding the Financial Crisis: Investment, Risk and Governance. SimCorp, 2009.

Welch, J. and Welch, S. (2006) Paying big-time for failure: Who is really to blame for those fat severance packages given to CEO's who falter? *Business Week,* 10 April 2006.

Williamson, O.E. (1964), The Economics of Discretionary Behavior: Managerial Objectives in a Theory of the Firm, Prentice-Hall, Englewood Cliffs, NJ.

About the author

Jean Dermine is Docteur es Sciences Economiques from the Université Catholique de Louvain and MBA from Cornell University and Professor of Banking and Finance at INSEAD, Fontainebleau. Author of articles on European banking markets, asset and liability management and the theory of banking, he has published five books, among which 'Bank Valuation and Value-based Management – deposit and loan pricing, performance evaluation and risk management', McGraw-Hill, is due for publication autumn 2009. He has been a visiting professor at Lausanne, Louvain, New York University, the Stockholm School of Economics, and the Wharton School.
Contact: jean.dermine@insead.edu

7
Avoiding international financial crises: an incomplete reform agenda

ABSTRACT In this paper, we argue that the set of reforms proposed recently by national and international groups to reduce the risk of future global banking crises constitutes an incomplete reform agenda. Three important issues remain, in our opinion, to be addressed:
- the accountability of bank supervisors and regulators;
- an end to the too-big-to fail doctrine;
- a satisfactory regime for the supervision and bailing out of international financial groups.

To address these issues, we review first briefly the origins of banking and the factors that contributed to the sub-prime crisis. This helps to evaluate the changes in regulation and financial market infrastructure proposed by various national and international institutions. It is then shown why these proposals, although useful, will not be sufficient to reduce significantly the risk of a future banking crisis.

7 Avoiding international financial crises: an incomplete reform agenda[1]

by Jean Dermine

The origin of the 2007 financial crisis and the reform agenda

To understand the roots of the 2007 financial crisis and the remedies proposed to reduce the likelihood of future crises, it helps to understand first the developments in banking markets, with the creation of the modern bank in Italy in the fifteen century, the interbank markets, securitisation, and the market for credit risk guarantees.

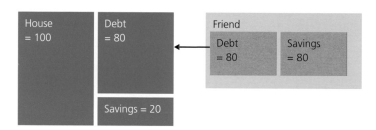

Fig. 1. *A world without banks*

The creation of banks

Let us travel back in time, three thousand years ago. There were no banks and no financial crises. As shown in figure 1, an investment in a house (a cabin) had to be financed out of one's personal wealth *(home equity)* or through a loan made by a friend. A drop in the value of the house by 10 would imply a drop in the value of home equity by 10 (figure 2). While allowing for transparency, the system was not efficient as it did not facilitate the mobilisation of savings and the financing of real investment. Modern banking in Europe was created in Italy in 1406. Banco di San Giorgio was founded in Genoa, several years before the founding in Siena in 1472 of the oldest surviving bank in the world, Banca Monte di Paschi.[2] As shown in figure 3, short-term deposits, attractive to savers, are invested in longer-term loans (attractive to borrowers). Imagine what the system would be, if financing the purchase of a house with a short-term loan, one would have to refinance the funding every year. This points to the maturity transformation role of banks, or, in modern economics, to the provision of liquidity insurance (Diamond-Dybvig, 1983; Allen and Gale, 2007). Whenever needed, deposi-

---

1 The author acknowledges the comments of S. Thomsen.
2 In March 2009, Monte dei Paschi received Italian government funding, after the ill-timed decision to buy Banca Antonveneta from the Spanish Santander in November 2007 (WSJ, 26/03/09).

tors can withdraw money.[3] In addition, banks perform a useful role in screening and monitoring borrowers. Diamond (1984) has shown that the diversification of risks reduces overall bank risk, the risk borne by depositors, and the cost of monitoring the bank. Since the creation of modern banks in the fifteenth century, the system was inherently fragile as depositors could run away, and illiquid loans could not be sold. For this reason, banks have been heavily regulated and safety nets, including emergency liquidity assistance by central banks or deposit insurance systems, have been created. Travellers in Brazil, South Africa or Russia will meet bankers who have experienced bank runs recently. The absence of major bank runs in the past 80 years in OECD countries is rather an exception. The observation that banking regulations and banking supervision have not been sufficient to prevent the 2007 banking crisis raises the question as to whether new regulations and supervisory models are needed or whether it is the enforcement of existing regulations that is the issue.

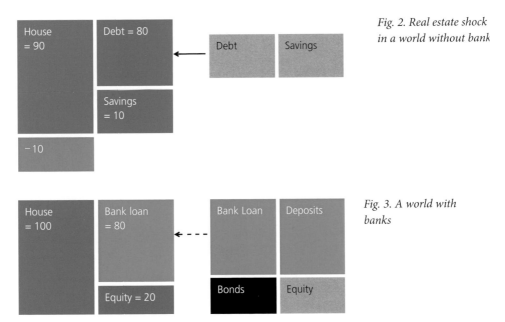

Fig. 2. Real estate shock in a world without bank

Fig. 3. A world with banks

The creation of interbank markets and securitisation

As demand for loans grew, banks short of cash started to borrow from other banks, often located in other countries. Such an example was the case of Icelandic banks borrowing from banks located in Germany. Another way to fund loans was to securitise them, that is to sell them to investors. This is described in figure 4. Due to asymmetric information problems – the bank knows more about the quality of loans than investors – this was never going to be an easy affair. Financial engineers issued asset backed securities that differ in seniority: senior tranche, junior or mezzanine tranches, and

---

3   A similar role of liquidity insurance is the offering of lines to credit, allowing borrowers to call money on demand.

equity tranche. The senior tranche was to a large extent protected from loan losses, which were covered by the junior and equity tranches. A question arises as to the size of junior and equity tranches needed to cover potential (unexpected) loan losses. Progress in statistics and credit modelling (Vasicek, 1987 and 2002) enabled the distribution of losses in a credit portfolio to be modelled. Finally, rating agencies were invited to rate these issues. Note that securitisation was not a completely new phenomenon, having been used by US government sponsored enterprises (GSEs), Fannie Mae and Freddie Mac, to securitise prime US mortgages for many years. What was new, in the period 2003-2007, was the securitisation of increasingly large classes of risky assets, including US sub-prime loans (Calomiris, 2007; Gorton, 2008). Securitisation reached $2 trillion in 2007, including 70% of recently issued US sub-prime loans.

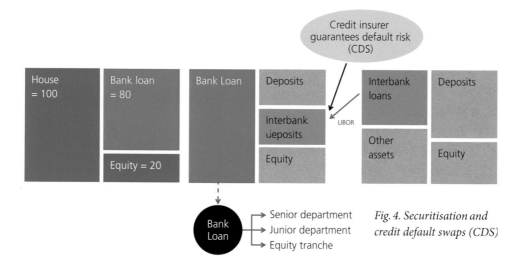

Fig. 4. Securitisation and credit default swaps (CDS)

Creation of credit default swaps (CDS)

A final type of financial innovation was the growth of a credit insurance market whereby one party would provide insurance coverage against the default (or downgrade) of another party. The market grew from $8 trillion in 2004 to $45 trillion in 2007. Looking at figure 4, it is useful to remember that financial markets (financial assets and claims) are just a mirror image of the real assets (real estate, industrial assets) in an economy. When the housing market falls by 10, it becomes difficult to estimate the impact on the value of bank's equity, on the value of a securitised asset, or on the value of a CDS. In other words, these credit transfer innovations have increased greatly the complexity and opacity of the market.

Apparently unnoticed by regulators was the fact that many of these securitisation vehicles, so-called structured investment vehicles (SIVs), involved the financing of long-term assets with short-term commercial paper (asset-backed commercial paper).

The plan was that this short-term paper would be rolled over at maturity. One notices a financial structure similar to that of a bank, with long-term assets funded with short-term funding. A major difference is that this short-term funding was highly volatile, as it is not protected by a public safety net. This new system has been dubbed quite correctly 'shadow banking'. In July 2007, the US investment bank Bear Stearns had to step in to refinance one of its flagship SIV, called High-Grade Structured Credit Strategies Enhanced Leverage Fund. A second issue was that the statistics on portfolio credit risk (Vasicek, 1987, 2002) requires the estimation of a critical variable, default correlation across loans. Little data on credit risk correlation at a time of a recession was available because the world economy had been expanding for the previous ten years. Moreover, data were nonexistent in a new untested market, the US sub-prime mortgage market. Finally, information on the counterparty risk exposure of financial institutions to the credit insurance market was, at best, incomplete. The lack of information on counterparty risk and therefore on the domino consequence of a default of a large institution seems to have been the reason for the bailout in September 2008 of the giant insurance group, AIG.

Given the emphasis of this article on the regulatory reform agenda, we will not address the forces explaining the explosion of these markets, nor shall we discuss the various actions taken by public authorities to prevent a collapse of their banking systems. Progress in the statistics of credit risk, a savings glut with large surplus in several countries (China and the Middle East in particular), a will of governments to facilitate home ownership, an accommodating monetary policy with exceptionally low interest rates, market pressures to grow income exercised on banks, asset managers and rating agencies, or simply greed and gullibility of investors are all likely to have contributed to the credit crisis (Hellwig, 2008; Greenspan, 2009). Following the bank run on Northern Rock in the United Kingdom in September 2007 (HM Treasury, 2009) and the default of Lehman Brothers and Washington Mutual (WaMu) in September 2008 in the United States, various measures were taken around the world to restore the stability of banks. Actions were taken on the asset side included insurance of credit risk or sale of toxic or legacy assets in the United States with the private-public investment programme (PPIP), and on the liability side with public guarantees of bank debt and capital infusion around the world.

As stated above, the objective of the paper is to discuss the regulatory and supervision framework, the financial market architecture needed to reduce the risk of financial crises in the future. Various reports have addressed these issues: the Turner Review (2009) by the Chairman of the Financial Services Authority (FSA) in the United Kingdom and the de Larosière Report (2009) for the European Union. Others include a report by the Group of Thirty (2009) (the Volcker report), reports by the Financial Stability Forum (2008, 2009) and two communiqués by the Group of 20 (2008, 2009). A list of regulatory proposals, common to all these reports, includes regulation of liquidity risk, market (trading) risk, capital, counterparty risk, compensation schemes, systemic risk and systemic institutions, and the control of cross-border financial institutions.

Control of liquidity risk
–   Control liquidity risk with stress test.

Control of market (trading) risks
–   Control market risk with stress test.
–   Increase capital regulation on illiquid assets.
–   Increase capital regulation on re-securitisation (securitisation of securitised tranches).

Control of capital
–   Increase tangible equity and core tier 1 relative to tier 2 securities. Impose a leverage ratio (equity to assets).
–   Reduce the pro-cyclicality of capital regulation, interpreted as capital regulations becoming binding at a time of a recession, with as a consequence a reduction of lending activity which itself exacerbates a recession. In reverse, capital regulation is not binding at a time of economic expansion, which facilitates lending, increasing the economic expansion and, possibly, bubbles in some asset markets. Proposal to create dynamic provision, to move away from marking-to-market of some assets, or to create a capital buffer in periods of economic expansion.[4]

Control of counterparty risk
–   Reduce counterparty risk by moving the CDS market to organised exchanges with margin calls and daily settlements.

Control of compensation schemes
–   Design financial incentives away from short-term profit to longer-term profit. Include clawback clauses in compensation schemes to introduce symmetry in the sharing of gains and losses.

Control of systemic risk
–   Control of macro systemic risk. The view is that the micro control of the soundness of individual financial institutions is not sufficient, and that public authorities

---

4   On the issue of dynamic loan loss provision and capital, some clarity appears to be needed. Provisions should be set to evaluate at the best the fair value of assets (taking into account probability of default, loss-given default). The objective is to evaluate the net economic value of a financial institution. Capital should be set to cover unexpected losses, to keep the net economic value positive, that is to ensure the survival and lending capacity of the bank in periods of severe recession. One would like to avoid the manipulation of provisions which reduce transparency and accountability of management. In Dermine (2008), I argue that too high capital regulation can have the unintended effect of forcing banks to divest of safe assets with too low margin.

need to control systemic risk, a risk rarely defined. This article interprets systemic risk as the threat of economic shocks that could affect jointly, several institutions, increasing correlation across the default risk of financial institutions. An example of a shock is a bubble in the real estate market or the devaluation of a currency when domestic units are running a foreign exchange risk. The de Larosière report (2009) proposes to replace the Banking Supervisory Committee, a college of European supervisors who meet at the European Central Bank),[5] by a European Systemic Risk Council (ESCR) presided by the President of the European Central Bank.[6] The G20 meeting in London in April 2009 proposed changes to the Financial Stability Forum, a club of Finance ministers, into a financial stability authority, charged with monitoring systemic risk with the International Monetary Fund.
– Closer supervision for systemic institutions.

Control of cross-border financial institutions
– Coordinate better the regulation and supervision of international financial groups. The de Larosière Report proposes to move progressively towards a European System of Financial Supervision (ESFS), with three committees[7] receiving more authority to arbitrage problems with the supervision of international groups. The G20 calls for creation of colleges of national supervisors to coordinate the supervision of international groups.

Additional measures proposed by the G20 are a curbing of tax heavens and the supervision of hedge funds that pose systemic risk, although neither of these seems to have been at the origin of the current crisis.

This long catalogue of measures to strengthen bank regulation and supervision is no doubt useful. But, one cannot refrain from asking two questions: aren't these proposals addressing issues of the past, observed during the crisis? Why were these measures not taken before the crisis, as it seems evident that very few institutions saw the crisis coming? A few quotes from reports published before and during the crisis[8] by the European Central Bank, the International Monetary Fund and the Basel Committee of Banking Supervision (BCBS) are telling:

November 2006 (report by the Banking Supervision Committee (BSC) of the European System of Central Banks (ESCB), (ECB, 2006, p4): "Vulnerability of the (banking) sector to adverse disturbances has diminished considerably."

---

5 Dermine, 2006.
6 Note that according to ECB (2007b), the Banking Supervisory Committee (BSC) was already 'contributing to the macro-prudential and structural monitoring of the EU financial system'.
7 With reference to the Lamfalussy Process (Dermine, 2006), three committees, such as the committee of European banking supervisors (CEBS), advise the European commission with steps to harmonise European banking regulations.
8 Public information on the risk of a crisis on the US sub-prime market can be dated to February 2007 with the announcement by HSBC of large loan losses in its US consumer finance subsidiary, Household International.

November 2007 (report by the Banking Supervision Committee (BSC) of the Europe-
an System of Central Banks (ESCB) (ECB, 2007): "Solvency should remain robust" (p7),
"financial positions of large EU banks remained strong in the first half of 2007" (p11),
"Tier 1 capital ratios remain adequate to cope with unexpected losses" (p16), "worsening
earning prospects, but shock absorption capacity remains comfortable" (p37).

The 2009 Turner Review in the United Kingdom (Turner Review, 2009) quotes the
International Monetary Fund (Global Stability Report, April 2006, IMF):

"There is a growing recognition that the dispersion of credit risk by banks to a broad-
er and more diverse group of investors, rather than warehousing such risk on their bal-
ance sheets, has helped make the banking system and overall financial system more
resilient.

The improved resilience may be seen in fewer bank failures and more consistent
credit provision. Consequently the commercial banks may be less vulnerable today to
credit or economic shocks."

Following the crisis, as discussed above, emphasis has been placed on strengthening regulations and supervision, such as those on liquidity risk and stress testing.
The Basel II capital regulations have been criticised for focusing too much on a single
capital measure (Acharya and Richardson, 2009). However, a reading of Pillar 2 of the
Basel II regulation finalised in June 2004 (BCBS, 2004) indicates explicitly that banking supervision must go beyond capital regulation:[9]

"# 738. Market risk: The assessment is based largely on the bank's own measure of
Value at Risk or the standardised approach for market risk. Emphasis should also be
placed on the institution performing stress testing in evaluating the capital to support
the trading function." (BCBS, 2004, p161).

"# 741 Liquidity risk: Liquidity is crucial to the ongoing viability of any banking
organisation. Banks' capital positions can have an effect on their ability to obtain li-
quidity, especially in a crisis. Each bank must have adequate systems for measuring,
monitoring and controlling liquidity risk. Banks should evaluate the adequacy of capi-
tal given their own liquidity profile and the liquidity of the markets in which they oper-
ate." (BCBS, 2004, p161).

One cannot but observe that Basel II was asking explicitly for stress testing in
market risk and liquidity risk. Why were these regulations not enforced? All the above
proposals deal with emphasis on existing regulations or on new regulations, but they
fail to address a central issue: the enforcement of regulations and conservative bank
supervision. Moreover, the emphasis is almost exclusively on a reinforcement of public
controls. No mechanism was suggested to develop private controls. Therefore, three
additional issues have to be addressed in a satisfactory manner if one wants to reduce
significantly the risk of future global financial crises:

9   Basel II regulations were applied in the European Union in 2008.

1   The accountability of banking supervisors. Banks were heavily regulated for many years. Why has the current supervision system allowed this crisis to happen?

2   The too-big-to-fail doctrine. There is an accepted view that large, systemic institutions should not be allowed to fail. This creates a problem as the credit risk on these institutions is not priced properly by the market, creating a risk of moral hazard. As an outcome of the resolution of the crisis is creating even bigger institutions (such as the merger of Bank of America, Countrywide and Merrill Lynch, or BNP-Paribas and Fortis), one needs to review this too-big-to-fail doctrine.

3   The supervision of international banking groups needs to be clarified.

These issues are discussed in the next section.

Three proposals to complete the reform agenda

Accountability of banking supervisors

It is a common observation that over the last two years, unlike the case of CEOs of banks, very few heads of national supervisory authorities have been invited to step down.[10] This raises a question of the accountability and independence of banking supervisors. An illustrative example is the case of several countries of Central and Eastern Europe (ECB, 2006, 2007a). Banks were allowed to lend massively in foreign currency (mostly Swiss francs, euro, and Yen) on the individual mortgage market. This created a large source of systemic risk, as the devaluation of the local currency would raise the rate of default in the entire banking system. Why has this source of systemic risk been allowed to develop?

It is not difficult to imagine that, for many ministers, a strong development in the real estate market helped the economy, employment, and the public budget with increased tax receipts. One would have needed a very brave bank supervisor to put a break on foreign currency lending, slowing down the economy and hurting real estate developers.

Another example is the risk division index imposed on mutual funds in some countries. So as to protect investors and provide them with minimum risk diversification, a mutual fund cannot invest more than 10% of its assets with the same counterparty. However, when money market funds were created in Europe, competing with bank deposits, banks lobbied successfully in some countries for an exception to this 10% rule. In Belgium for example, money market funds can invest up to 25% of assets in money market certificates issued by the sponsoring bank (CBFA, 1991). For the sponsoring bank, the exception to the 10% risk rule was welcomed as it enabled them to retain liquidity in the bank. The argument for the exception is that, regulated by

---

10   Exceptions include the head of banking supervision in Ireland, and the president of the central bank in Iceland.

competent authorities, the risk of bank default is minimal. This is an example of how lobbying and a twist in regulation may harm the systemic stability of the financial system if worries about bank soundness start to migrate to the money market funds market. Some (Rochet, 2008) are calling for truly independent banking supervisors who would be evaluated on their success in maintaining the stability and soundness of a banking system. In the same way as central banks have been given independence to conduct monetary policy to control inflation, independence would be given to banking supervisors to ensure soundness in the banking system. One must recall that the independence of central banks is justified by the argument that, due to short cycles of political elections, ministers of finance will be tempted to expand money supply to steer the economy in the short term, at a cost of creating inflation and longer term ills for the economy. To reduce the impact of lobbying or too short-term goals of public policy-makers, independence was given to central banks to control inflation. Should banking supervision be given similar independence?

The de Larosière report (2009, footnote 10, p 47) is very explicit on this issue and the limited degree of independence that should be given to banking supervisors:

"As such the supervisory authority must be empowered and able to make its own independent judgements (for instance with respect to licensing, on-site inspection, off-site monitoring, sanctioning, and enforcement of the sanctions), without authorities of the industry having the right or possibility to intervene. Moreover, the supervisor itself must base its decision on purely objective and non-discriminatory grounds. However, supervisory independence differs from central banking independence (that is, in relation to monetary policy), in the sense that the government (usually the Finance Minister) remains politically responsible for maintaining the stability of the banking system, and the failure of one or more financial institutions, markets or infrastructure can have serious implications for the economy and the tax payer's money."

The de Larosière report is very explicit on the limited independence of banking supervisors. It says that day-to-day banking supervision should be independent but, unlike monetary policy, the government is responsible for maintaining the stability of the banking system. Since this system seems to be prone to regulatory capture and/or short-term political interest, one wonders whether banking supervision should not receive the same degree of independence as that given to monetary policy-makers and central banks. Note that the decision to bail out an institution with implications for taxpayers' money would still remain with the government. What would be given to the independent authority is the power to legislate and to enforce supervision with the sole objective of preserving the stability and soundness of the banking system. In the author's opinion, given that it is not a lack of regulations but poor enforcement by banking supervisors that has contributed to the crisis, a reinforcement of the accountability of banking supervisors is a must. Independence with clearly set objectives among which would be the soundness of the banking system, would contribute greatly to accountability.

The 'too-big-to-fail' doctrine

Several of the international proposals call for closer supervision of systemic institutions. Although not explicitly mentioned, as central banks have generally chosen to keep some ambiguity, there is an implicit recognition that the default of large institutions would cause systemic risk, and that, in case of distress, they would benefit *de facto* from some type of public support. As public support has often taken the form of public guarantees on the debt issued by these firms, the systemic institutions benefit from a significant source of competitive advantage: the ability to raise debt at a lower cost. There is no doubt, in the author's opinion, that this state of affairs has facilitated the creation of very large financial institutions. And one is forced to recognise that the resolution of the crisis has created even larger systemic institutions. In the United States, Bank of America has purchased Countrywide and Merrill Lynch. In Europe, the Belgian government has sold Fortis Bank to the French BNP-Paribas. In the United Kingdom, Lloyds Banking Group has been formed with the merger of Lloyds TSB and HBOS. To resolve the moral hazard resulting from the implicit debt guarantee, the authorities are calling for closer supervision of systemic institutions with potentially a larger capital ratio, as, unregulated, they would not take into account the systemic cost of a failure.

Again, we recognise the need for close supervision of systemic institutions. But as these institutions have been identified long ago (Dermine, 2000), one is left wondering again why large institutions, such as Citigroup or UBS, were allowed by supervisors to grow so much and take large risks. The previous proposal, granting more independence to supervisors, would reinforce their power and accountability. A complementary mechanism would be to increase private discipline. Several participants have claimed that private discipline has functioned properly during the crisis as shareholders of banks have been penalised dearly. But, this is not enough. Equity is a very small fraction of the funding of banks' assets. In most cases so far, bank debt, including non-insured deposits, interbank debt, subordinated debt, has been protected, often with explicit public guarantees. A way to increase scrutiny of banks, beyond that of supervisors, is to put bank debt at risk, that is to create the risk of bank failure. As these systemic institutions are vital for the proper functioning of the economy, one needs to design a bankruptcy system that allows for the benefits of default (increase in market discipline), with a maximum reduction of the cost of default (Dermine, 2003, 2006).

To reduce the cost of the default of a large bank, three features must be met. The first is that the bank should be closed for only a few days (during a weekend). As depositors and consumers need to access their funds rapidly, the resolution of bankruptcy must be swift. Bankruptcy court must be able to swap debt for equity very rapidly before the re-opening of the bank. Secondly, to avoid the fear of domino effects, the failure of one bank causes the failures of others, credible information on counterparty risk must be available on the spot. Third, to reduce the risk of a bank run by uninsured short-term depositors and instability, a mechanism must be put in place to prevent capital flight attempting to avoid the default of a bank. Having no place to hide, they will have no incentive to run (Baltensperger and Dermine, 1986). In short, all banks should be able to meet the bankruptcy acid test: they can be put into bankruptcy. If it is not feasible, then the structure of the financial institution must be changed. With bank debt at risk, there will be much more pressure from private financial markets to limit bank risk. The current proposal calling for closer supervision of systemic banks could be a step in the wrong direction if it increases incentives for banks to become large and systemic.

Finally, a last recommendation on how to complete the international reform agenda, is to address more completely the supervision and corporate structure of international financial institutions.

The supervision of international banking groups

Both the G20 communiqués (2008, 2009) and the de Larosière Report (2009) have highlighted the need for an adequate supervision of international banking groups. The call is to create effective colleges of national supervisors to ensure better supervision of international banking groups. In the United Kingdom, the *Turner Review* (2009) also emphasises a proper regulation of domestic and international banks operating in the UK. This followed the collapse of Icelandic banks which collected deposits through a branch network in the UK. Note that an adequate supervision of international banks operating abroad with branches or subsidiaries is not a new issue. The collapse of the German Bank Herstatt in 1974 prompted the *Basel Concordat* (BCBS, 1983). It called, in the case of branches, for regulation of capital by the home regulator of the parent, and liquidity regulation by the host state. Later on, the defaults of Banco Ambrosiano Holdings and BCCI again drew attention to the supervision of international groups. Recent calls to improve the supervision of international financial firms confirm that lessons of history have not yet delivered satisfactory responses. In the European Union, although the single market has been in place since 1992, the creation of significant cross-border banking groups is a recent phenomenon (Schoenmaker and Osterloo, 2004). It started with the creation of Nordea, the merger of four Nordic banks completed in 2000, the acquisition of Abbey National by Banco Santander in 2004, followed by Unicredit and HVB in 2005, and BNP-Paribas with BNL (2006) and Fortis (2009).

In the context of cross-border banking, one needs first to identify two specific issues These concern, successively: (1) the presence of cross-border spillover effects and (2) the financial ability of some countries to deal with bailout costs and large and complex financial institutions.[11] As the European Union is home to more advanced

---

11   A third issue of a more technical nature is deposit insurance coverage and the ability to reimburse depositors rapidly in case of bank closure (Dermine, 2006).

cases of international banking integration, a discussion of the adequacy of current institutional structure follows. We then draw conclusions on the G20 world level proposals.

In the European Union, the supervision of a bank operating through a branch network is supervised by the parent home supervisor, while a subsidiary is controlled by the host authority.

Two specific issues of international banking

The first and main issue concerns cross-border spillovers and the fear that the provision of financial stability, a public good, by national authorities might not be optimal. Four types of potential cross-border spillovers can be identified: i. cross-border cost of closure, ii. cross-border effects of shocks to banks' equity, iii. cross-border transfer of assets, iv. cross-border impact on the value of the deposit insurance liability.

i. Cross-border cost of closure

Imagine that a foreign bank buys a Dutch bank, and converts it into a branch. According to EU rules on home country control, the Dutch branch would be supervised by the authority of the foreign parent.[12] However, Dutch authorities remain in charge of financial stability in the Netherlands. The Dutch treasury could be forced to bail it out for reasons of internal stability, but would not have the right to supervise the branch of a foreign bank because of home country control. Since the lender-of-last-resort and the treasury will be concerned primarily with their domestic markets and banks operating domestically,[13] and since they will bear the costs of a bailout, it is legitimate for insurers to retain some supervisory power over all institutions (branches and subsidiaries) operating domestically. That is, host country regulations could apply to limit the risks taken by financial institutions and the exposure of the domestic central bank or treasury in cases of bail out.[14] In other words, home country control has to be complemented by some form of host control as long as the cost of bailout remains domestic. This position appears to have been partly recognised by the European Commission which states that "in emergency situations the host-country supervisor may, subject to ex-post Commission control, take any precautionary measures" (Walkner and Raes, 2005, p.37). This argument is again present in the British *Turner Review* calling for effective supervision of all institutions operating in the UK. In this case, since the default of a large international bank could affect several countries, the optimal decision to bail out should take into account the interest of all parties-countries

---

12   Although Iceland is not a Member of the European Union, the provisions of the single market applies to it due to its membership in the European Economic Area (EEA). This is the reason why branches of Icelandic banks were allowed to operate in the UK with apparently minimum supervision by the British authorities (Turner Review, 2009).

13   It is well known that the Bank of Italy did not intervene to prevent the collapse of the Luxembourg-based Banco Ambrosiano Holding, because it created little disruption on the Italian financial markets.

14   Bailing out would occur if the failure of a branch of a foreign bank led to a run on domestic banks.

affected. As is discussed below, such a decision could be transferred to the European level or should at least require coordination among these countries.

ii. Cross-border effects of shocks to banks' equity

Peek and Rosengren (2000) demonstrated the impact on the real US economy of a drop in the equity of Japanese banks, resulting from the Japanese stock market collapse in the 1990's. The transmission channel runs through a reduction in the supply of bank credit. Since in a branch-multinational bank, the home country will control solvency through policy on loan-loss provisioning and validation of probability of default (PD) in the Basel II framework, it could have an impact on the real economy of the foreign country.[15] In the current crisis, a similar fear has been expressed by countries in Central and Eastern Europe having banking systems controlled largely by foreign banks.

iii. Cross-border effects of transfer of assets

In a subsidiary-type multinational, in which the host country retains supervision of the subsidiary, there could be a risk that the home country colludes with the parent bank to transfer assets to the parent bank. This risk has been discussed in the countries of Central and Eastern Europe. An example is the case of Lehman Brothers Holdings. The investment bank had a policy of centralising cash operations in New York. In the case of Lehman Europe, the 'cash sweep' had taken $8 billion out of the UK business the Friday before it collapsed.[16]

iv. Cross-border effects on deposit insurance

The general argument is related to diversification of risks in a branch-based multinational, which, because of co-insurance, reduces the value of the put option granted by deposit insurance (Repullo 2001, Dermine 2003). There is an additional dynamic consideration to take into account. A multinational bank could be pleased with its overall degree of diversification, while each subsidiary could become very specialised in local credit risk. This implies that banks in a given country could find themselves increasingly vulnerable to idiosyncratic shock. One could argue that, for reasons of reputation, the parent company would systematically bail out the subsidiaries as if they were branches. This could be true in many cases, but there will be cases where the balance of financial costs versus reputation costs may not be so favourable. Capital adequacy rules should be less stringent on branch-based structures. The more general point is that diversification should be rewarded with less stringent capital adequacy ratio.

Cross-border spillovers raise the question of whether coordination of national in-

---

15 The Basel Committee on Banking Supervision (2003) discusses the respective role of host and home country in validation of PDs. It calls for adequate cooperation between host and home authorities, and a lead role for the home country authority. In the case of Nordea, the Swedish supervisory authority will have the final control of the measurement of PD across the group. This is in line with recommendation by the European Union (ECB, 2007b).

16 *Financial Times,* 15 April 2009.

terventions will be optimal (Freixas, 2003). This will be discussed in the context of an assessment of the European institutional architecture.

Bailing out costs: Too big? Too complex?

A second issue is that the bailout of a large bank could create a very large burden for the treasury or deposit insurance system of a single country (Dermine 2000, 2006). This was precisely the case of the Icelandic banks whose size exceeded vastly the insurance capability of the country. A related issue that has received much interest is that some banks have become too complex to fail. Imagine the case of a large European bank with significant cross-border activities and non-bank activities, such as insurance or asset management, which runs into financial distress. It would be very difficult to put such a bank into receivership. Given the complex web of corporate subsidiaries and the various legal complexities, the uncertainty concerning the costs of a default is likely to be high, and this complexity might create a temptation for a bailout ('too big and too complex to fail').

Two cross-border banking issues related to financial stability have been analysed: cross-border spillover effects and the size and complexity of bailout. Let us now review the adequacy of the current EU institutional structure.

Adequacy of EU institutions

The EU institutional structure currently in place to deal with financial crises has received a great deal of attention in the last ten years (ECB, 2007b). A series of committees have been created to facilitate cooperation and exchange of information, and a directive on winding up financial institutions has been adopted.[17] The Brouwer reports (Economic and Financial Committee, 2000 and 2001) had very much validated the current EU institutional structure to deal with a financial crisis. They essentially argued that there would be no legal impediment to the transfer of information across borders, and recommended an additional effort to strengthen cooperation through a Memorandum of Understanding (MOU) dealing with crisis situations.[18] In May 2005, it was announced that an emergency plan for dealing with a financial crisis had been agreed by the European Union finance ministers, central bankers and financial regulators.

---

17   There are currently four potential forums for coordination. The *European Banking Committee* assists the European Commission in preparing new banking community legislation. The *Committee of European Banking Supervisors (CEBS)* is concerned with the application of EU regulations. At the *EU Groupe de Contact (GdC)*, national supervisors of banks meet regularly to exchange information. At the European Central Bank, the *Banking Supervisory Committee (BSC)* works in the context of the Eurosystem's task of contributing to the smooth conduct of polices pursued by the competent national authorities relating to the supervision of credit institutions and the stability of the financial system (Article 105 (5) of the Treaty on European Union).

18   One cannot fail to notice the excess of optimism in these reports. Indeed, the 2009 de Larosière reports the lack of resources and authority of the Committee of European Banking Supervisors (CEBS) and the inadequate cooperation in the case of handling the crisis of the Belgian-Dutch Fortis (despite the existence of a MOU between the Dutch and Belgian authorities, ECB, 2007b).

A MOU among the 25 EU members which facilitated the exchange of information was tested in 2005 and 2006 with a full-scale simulation of a financial crisis.[19] But, the ECB opposed a move to agree on *ex ante* sharing rules for financing of a bailout.[20] The latest MOU on cooperation between the financial supervisory authorities, central banks and financial ministries of the European Union was signed on 1 June 2008. Signatories include 58 financial supervisory authorities, 28 central banks including the ECB, and 28 finance ministries (one for each of the 27 UE members, with the exception of Denmark which has a ministry of finance and a ministry of economics as signatories). The number of signatories (114 in total) is representative of the potential complexity in dealing with large international groups in the European Union. The 2008 MOU calls for Voluntary Specific Cooperation Agreement (VSCA) on crisis management and resolution between the finance ministries, central banks and financial supervision authorities of countries A, B and C etc.

In the context of the Financial Services Action Plan, the directive on 'Winding up of Credit Institutions'[21] was adopted in 2004, 16 years after it was first proposed. It is consistent with the home regulator principle. When a credit institution with branches in other member states fails, the winding up process will be subject to the single bankruptcy proceedings of the home country. Note that, although recognised as a significant piece of legislation to avoid the complexity issue, it falls short of solving the subsidiaries issue.

Accepting the accountability principle, according to which banking supervision, deposit insurance, and bailing out should be allocated to the same country, it appears that, in European banking, there are three ways to allocate banking supervision: to the 'host' state, to the 'home country', or to a European entity. The pros and cons or the three approaches are reviewed.

In the 'host' state approach, adopted in New Zealand,[22] multinational banks operate with a subsidiary structure. The national central bank retains control of banking supervision, deposit insurance, and bailout.[23] This system suffers from three drawbacks. First, it does not allow banks to realise fully the operating benefits expected from branch banking (Dermine, 2003 and 2006). Second, a subsidiary structure contributes to the creation of large and complex financial institution (LCFI). Subject to different bankruptcy proceedings, the closure of a large international bank would become very complex. In a branch structure, the European directive on winding up would be applicable, subjecting the bankruptcy proceedings to one country. Third, the resolution of a crisis could be hampered, as discussed above, by problems linked

---

19 Financial Times 16 May 2005 and 10 April 2006. ECB (2007,b).
20 Financial Times 23 April 2007 and 17 September 2007; ECB (2007b).
21 Official Journal 125, 05.05.2001.
22 According to Schoenmaker and Oosterloo (2004), to allow them to exercise control, authorities in New Zealand require some overseas banks to establish locally incorporated subsidiaries, instead of operating as branches.
23 It must be observed that the 'host country' approach is applied as well in Europe when banks operate abroad with subsidiaries.

to transfer of assets from subsidiaries to the parent, or to problems of sharing of information. It appears that a 'host country'-based system would not allow full realisation of the expected benefits of European integration.

The second system, the 'home country' approach, currently applicable to cross-border branches in the European Union, suffers from two drawbacks. The first is that small European countries, such as Ireland, Sweden, Belgium, Switzerland, or the Netherlands, may find it difficult to bear the cost of the bailout of a large international bank.[24] European funding might be needed. The second is related to the cross-border spillover effects. The decision to close a bank could affect other countries. In principle, cooperation among countries could take place in such a situation, but one can easily imagine that conflicts of interest between countries on the decision to close a bank will arise, and that the sharing of the bailing out costs among countries will not be simple (Schoenmaker and Oosterloo, 2004; Schoenmaker, 2009). These conflicts of interest could, at times, even limit the cross-border exchange of information among regulators. Freixas (2003) have called for *ex ante* rules to force this transfer of information. Goodhart and Schoenmaker (2009) explore *ex ante* mechanisms for fiscal burden sharing in a banking crisis in Europe

The de Larosière Report is the first serious call on the lack of adequacy of current arrangement to prevent and deal with financial crises. Besides the replacement of the Banking Supervisory Committee (BSC) by a European Systemic Risk Council (ESRC), the Report proposes to move gradually to a European System of Financial Supervision (ESFS). This is in line with the G20 proposal of creating a college of national banking supervisors to supervise cross-border banks, but it goes further. It wants to give authority to the Committee of European Banking Supervisors to take decision and arbitrage eventual conflicts between national supervisors. In a first stage (2009-2010), the CEBS would be transformed into a European Banking Authority. In a second stage, the European System of Financial Supervision would be created. It would include the college of national supervisors, the newly created European Banking Authority and observers from the ECB/ESCB.

In our opinion, the de Larosière proposal does not go far enough. An adequate public architecture for the proper functioning of financial markets must deal with three issues: regulation, supervision, and the decision to bail out, including financing with taxpayers money. The creation of the ESFS is dealing with the first two issues, regulation and supervision. Instead of creating a European body, it relies on a college of supervisors and a banking authority that can arbitrage when necessary. Time will tell if this can function or if there is a need to integrate this network into a single European institution. As concerns the last issue, the decision to bail out, it is left to the

---

24   Note that large European countries might prefer the *status quo* to make the expansion of banks from
     smaller countries more difficult. One observes that the resolution of the banking crisis with public funding
     has increased the public debt of small countries, with a significant impact on the spread paid on this debt.

national level. As such, it does not solve the problem of externality discussed earlier. It is therefore not a surprise that, although calling for international cooperation, the British *Turner Review* wants to strengthen host country supervision in the United Kingdom.

The lesson of the European Union with colleges of supervisors and memorandum of understanding is that it does not work satisfactorily at a time of a crisis. The G20 call for the creation of colleges of national supervisors ignores this EU experience. In the European Union, although progress is made in the de Larosière Report in dealing with the first two issues, regulation and supervision, a gap still exists with the bailing out decision. The severity of the international crisis calls for more advanced solutions to cross-border banking.

The 'home country' approach has served European banking well until now because the scale of cross-border banking was limited and banks were operating abroad mostly with subsidiaries. Host country supervision was effectively taking place. The crisis has highlighted two facts: first that branch banking was starting to take place, limiting control by the host authority. Secondly, that in case of failure, international cooperation was not effective. The two drawbacks of the 'home country' approach, discussed above, lead us to call again for a European-based system of banking supervision and deposit insurance for large international banks. An international bank would be defined either by its size, equity relative to the GDP of one country (say, 3%), or by its market share in a foreign country (say 10%). Goodhart (2003) argues that a European supervisory agency cannot exist as long as the cost of the bailing out is borne by domestic authorities, with reference to a British saying 'He who pays the piper calls the tune'. There is no disagreement with this accountability principle, but the recommendation to move banking supervision, deposit insurance, and bailing out to the EU is motivated first by the fact that, de facto, the bailout of large banks from small countries will be borne by European taxpayers, and, second, that spillover effects demand a coordinated resolution. A discussion of cross-border bailout will turn rapidly into an issue of taxpayers' money and into a constitutional debate. A discussion on the preference of citizens to define the border of the nation at the country or European level cannot and should not be avoided. It will guide the choice among 'host country' or EU-wide control of international banks.

At world level, it seems that very strong host supervision should prevail. This is very much the case as international banking often develops with subsidiaries (Dermine, 2003).

Conclusion

Following the international banking crisis, a series of proposals have been made to reinforce the quality of banking supervision. They concern: liquidity, trading risk, capital and provisioning, compensation policies, systemic risk, and college of national supervisors. It is the author's opinion that, although useful, these proposals stop short of addressing three main issues: how to increase the accountability of supervisors; how to limit the too-big-to-fail doctrine and how to supervise international groups.

It is striking to observe that pre-crisis regulation, such as Basel II, was already calling for stress tests on liquidity and market risk. The supervision issue in the crisis was not a lack of regulation but a lack of enforcement of regulation and effective supervision. How to increase accountability of bank supervision seems to be the issue. The second issue is the existence of systemic banks. Their existence seems to be taken for granted, with a need to guarantee their debt and, as a consequence, to supervise them closely to limit moral hazard. An alternative would be to design an adequate bankruptcy process for these institutions. The fear of bankruptcy would raise private incentives of debt holders to monitor bank risks. Finally, the G20 proposal of a college of supervisors will not help much if one reads the last 10-year experience of the European Union. The 'home country' principle has served European banking integration well until now because most banks were operating across borders with subsidiaries (regulated by host authorities). Cross-border banking with branches and spillover effects caused by bank failure call for European cooperation. And, in addition, small European countries may not be able, as the extreme case of Iceland illustrates, to bail out their large international banks. This raises the issue of transferring the cost of bailing out, deposit insurance, and banking supervision of large international banks to a European entity. Current policy proposals stop short of this eminently political step.

References

Acharya, V.V. and M. Richardson (2009): 'Restoring Financial Stability', Wiley Finance.

Allen, Franklin and Douglas Gale (2007): 'Understanding Financial Crises', Oxford University Press.

Baltensperger, Ernst and Jean Dermine (1987): 'Banking Deregulation in Europe', *Economic Policy*, 4, 1987.

Basel Committee of Banking Supervision (1983): 'Principles for the supervision of banks' foreign establishments (Basel Concordat).

Basel Committee of Banking Supervision (2004): 'International Convergence of Capital Measurement and Capital Standards' (comprehensive version published in June 2006), Bank for International Settlements, Basel.

de Larosière, Jacques (2009): 'The High-Level Group on Financial Supervision in the EU', European Commission, Brussels, 1–85.

Calomiris, Charles W. (2007): 'The Subprime Turmoil: What's old, What's New, and What's Next? October, Federal Reserve Bank of Kansas City Jackson Hole Conference, 1–112.

Commission Bancaire, Financière et Assurance (CBFA)(1991): Arrêté Royal Relatif à Certains Organismes de Placement Collectif, Article 56, 4 Mars, Brussels.

Dermine, Jean (2000): 'Bank Mergers in Europe, the Public Policy Issues', *Journal of Common Market Studies,* 38 (3), September.

Dermine, Jean (2003): 'European Banking, Past, Present, and Future', in *The Transformation of the European Financial System* (Second ECB Central Banking Conference), eds V. Gaspar, P. Hartmann, and O Sleijpen, ECB, Frankfurt.

Dermine, Jean (2006): 'European Banking Integration, Don't Put the Cart before the Horse', *(Journal of) Financial Markets, Institutions and Instruments,* Vo. 15 (2) 2006.

Dermine, Jean (2008): A Blueprint for a New International Financial Order', INSEAD 'Knowledge (http://knowledge.insead.edu/NewFinancialOrder081104.cfm).

Diamond, Douglas W. (1984): 'Financial Intermediation and Delegated Monitoring', *Review of Financial Studies,* 51, 393–414.

Diamond, Douglas and Phil Dybvig (1983): 'Bank Runs, Deposit Insurance and Liquidity', *Journal of Political Economy,* 91, 401–419.

Economic and Financial Committee (2000) : 'Brouwer Report on Financial Stability', Economic Papers 143, 1–33.

Economic and Financial Committee (2001): 'Report on Financial Crisis Management', Economic paper 156, 1–32.

European Central Bank (2006):'EU Banking Sector Stability', November.

European Central Bank (2007a):'EU Banking Sector Stability', November.

European Central Bank (2007b): 'The EU Arrangements for Financial Crisis Management', ECB Monthly Bulletin, February, 73–84.

European Union (2008): 'Memorandum of Understanding on Cooperation between the Financial Supervisory Authorities, Central Banks and Finance Ministries of the European Union', ECFIN/CEFCPE(2008)REP/53106REV REV, 1 June.

Financial Stability Authority (2009): The Turner Review, 'A Regulatory Response to the Global Banking Crisis', March.

Financial Stability Forum (2008): 'Report of the Financial Stability Forum on Enhancing Market and Institutional Resilience', April, 1–70.

Financial Stability Forum (2009): 'Report of the Financial Stability Forum on Enhancing Market and Institutional Resilience – Update on Implementation', April, 1–17.

Freixas, X. (2003): 'Crisis Management in Europe', in *Financial Supervision in Europe* eds J. Kremers, D. Schoenmaker, and P. Wierts, Cheltenham: Edward Elgar, 102–119.

G20 (2008): 'Summit Communique', 16 November, Washington D.C.

G20 (2009): 'Summit Communique', 2 April, London.

Goodhart, C. (2003): 'The Political Economy of Financial Harmonization in Europe', in *Financial Supervision in Europe* eds J. Kremers, D. Schoenmaker, and P. Wierts, Cheltenham: Edward Elgar, 129–138.

Goodhart, Charles and Dirk Schoenmaker (2009) : 'Fiscal Burden Sharing in Cross-Border Banking Crises', *International Journal of Central Banking,* forthcoming, 1–16.

Gorton, Gary (2008): 'The Panic of 2007', Federal Reserve Bank of Kansas City, Jackson Hole Conference, August, 1–91.

Greenspan, Alan (2009): 'The Fed Didn't Cause the Housing Bubble', *Wall Street Journal,* 12 March.

Group of Thirty (2009): 'Financial Reform, a Framework for Financial Stability', Washington DC, 1–70.

Hellwig, Martin (2008): 'Systemic Risk in the Financial sector: An Analysis of the Subprime-Mortgage Financial Crisis', Max Planck Institute for Research, November, 1–73.

HM Treasury (2009): 'The Nationalization of Northern Rock', National Audit Office, 20 March 2009, 1–58.

Peek, J. and E. Rosengren (2000): 'Collateral Damage: Effects of the Japanese Banking Crisis on Real Activity in the United States', *American Economic Review,* Vol. 90 (1), 30–45.

Repullo, R. (2001): 'A Model of Takeovers of Foreign Banks', *Spanish Economic Review,* 3, 1–21.

Rochet, Jean-Charles (2008): 'Why are there so many banking crises? The politics and policy of banking regulation', Princeton University Press.

Schoenmaker, Dirk (2008): 'The Trilemma of Financial Stability', VU University Amsterdam 1–21.

Schoenmaker, D. and S. Oosterloo (2004): 'Cross-border Issues in European Financial Supervision, forthcoming in D. Mayes and G. Wood eds. *The Structure of Financial Regulation,* London: Routledge.

Vasicek, Oldrich A.(1987): 'Probability of Loss on Loan Portfolio', Working Paper, KMV Corporation, 1–4.

Vasicek, Oldrich A. (2002): 'Loan Portfolio Value', *Risk,* December, 160–162.

Walkner, C. and J.P. Raes (2005): 'Integration and Consolidation in EU Banking, an Unfinished Business', European Economy # 226, April, 1–4.

About the author

Professor Ingo Walter, director of SimCorp StrategyLab, holds the Seymour Milstein Professorship of Finance, Corporate Governance and Ethics and serves as the Vice Dean of Faculty at the Stern School of Business, New York University. He has been on the faculty at New York University since 1970. He has had visiting professorial appointments at the Free University of Berlin, University of Mannheim, University of Zurich, University of Basel, the Institute for Southeast Asian Studies in Singapore, and various other academic and research institutions. He also held a joint appointment as Professor of International Management from 1986 to 2005 and remains a visiting professor at INSEAD in Fontainebleau, France.

In addition to his position as a director of SimCorp StrategyLab, Professor Walter has served as a consultant to various corporations, banks, government agencies and international institutions, and has held a number of board memberships. Contact: iwalter@stern.nyu.edu

8
Reputational risk and
the financial crisis

ABSTRACT In recent years the role of various types of financial intermediaries has evolved dramatically, as financial market deregulation and innovation have resulted in intensified competition among banks, asset managers, insurers and securities firms – each competing vigorously with traditional and non-traditional rivals. Consequently, market developments have periodically overtaken regulatory capabilities intended to promote financial stability and fairness as well as efficiency and innovation. The ongoing financial crisis is a prime example. It is unsurprising that these conditions would give rise to significant reputational risk exposure for the financial firms involved. This chapter focuses on defining, assessing and pricing reputational risk – a type of risk that is often associated with major losses in the going-concern value of financial intermediaries – that is, a major decline in the risk-adjusted value of expected future earnings as a result of reputation-sensitive events. The financial crisis has given rise to many such events, so that assessment of reputational risk has risen to the forefront of attention on the part of managements and boards.

8 Reputational risk and the financial crisis

by Ingo Walter

The global financial crisis of 2007–09 has been associated with an unprecedented degree of financial and economic damage. For investors and financial intermediaries, the estimates seem to have risen thus far to over $4 trillion or so worldwide by the time things have begun to stabilise, according to the International Monetary Fund (2009). Along with the financial damage has come substantial reputational damage, for the financial services industry, for financial intermediaries and asset managers and for individuals.

At the firm level, for example, the 2008 $150 billion bailout of AIG and its *de facto* nationalisation by the US government – attributable to massive losses in a single unit of the holding company (AIG Financial Products) – left the government on the horns of a dilemma. The value of AIG's insurance operating units, for the most part perfectly sound, had seen their reputations so badly tarnished by the disaster in the parent organisation's performance that clients had taken their business elsewhere and competitors were actively recruiting its staff, so that buyers willing to pay realistic prices for the businesses were few and far between. This made those businesses virtually impossible for the government to sell and thereby recoup taxpayer assistance to AIG, which in turn required further government financing while waiting for markets to turn around. AIG's reputational capital suffered further damage in a furore over executive bonuses, congressional hearings, legal subpoenas and jokes in the media. Insurance premium income dropped by 22% from 2007 to 2008. So waiting for higher prices for sound AIG insurance businesses risked further erosion of their franchise value in the meantime. With its name 'thoroughly wounded and disgraced,' AIG launched a major rebranding effort as American International Underwriters to try to rebuild its reputation.[1]

Crisis-driven reputational damage at the firm level can also be inferred from remarks by Peter Kurer, former Supervisory Board Chairman of UBS AG, who noted at the bank's annual general meeting in April 2008 that "We shouldn't fool ourselves. We can't pretend that there has been no reputational damage. Experience says it goes away after two or three years."[2] Perhaps it does, perhaps not, but the haemorrhage of private client withdrawals at the height of the crisis suggests severe reputational damage to the world's largest private bank. And the number of financial firms ranging from Santander in Spain to Citigroup in the US and Union Bancaire Privée in

---

1   Mary Williams Walsh, 'At A.I.G. The Brand is Tarnished', New York Times, 23 March 2009.
2   See http://careers.hereisthecity.com/front_office/corporate_and_investment_banking/press_
     releases/124cntns.

Switzerland that have reimbursed client losses from the sale of bankrupt Lehman bonds, collapsed auction-rate securities and investments in Bernard Madoff's fraudulent scheme suggests that firms are in fact highly sensitive to lasting reputation losses. And at the individual level the world is full of disgraced bankers whose hard work, career ambitions and future prospects lie in tatters.

At the industry level, on the other hand, Josef Ackermann, CEO of Deutsche Bank and Chairman of the International Institute of Finance, noted in April 2008 that the industry was guilty of poor risk management. This involved serious over-reliance on flawed models, inadequate stress-testing of portfolios, recurring conflicts of interest and lack of common sense as well as irrational compensation practices not linked to long-term profitability. There was a growing perception by the public of 'clever crooks and greedy fools' and he concluded that the industry had a great deal of work to do to regain its reputation.[3]

This raises the issue of 'systemic' damage to the financial intermediation industry during the crisis evolution. Since biblical times, 'bankers and moneychangers' have battled to build the reputation of their profession – a difficult task for people whose job may be to deny credit, collect money owed, or trade in various financial instruments in ways that inflict losses on others. When such activities threaten to destroy the financial system as a whole, the collective reputations of activities that comprise banking and finance may well be tarnished in a 'systemic' way, triggering draconian policy and regulatory actions that may very well damage an industry that is the fulcrum of capital allocation in the modern global economy.

Whether at the industry, firm or individual level, the reputational costs of the financial crisis have been enormous. Section 1 of this chapter considers the 'special' nature of financial services and traces the roots of the reputational risk that firms in the industry invariably encounter. Section 2 defines what reputational risk is, and outlines the sources of reputational risk facing financial services firms. Section 3 considers the key sources of reputational risk in the presence of transaction costs and imperfect information.[4] Section 4 surveys available empirical research on the impact of reputational losses imposed on financial intermediaries, including the separation of reputational losses from accounting losses. Section 5 considers managerial requisites for dealing with reputational risk issues, notably from a corporate governance perspective. Section 6 concludes with some governance and managerial implications.

The special character of financial services

Financial services comprise an array of 'special' businesses. They are special because they deal mainly with other people's money, and because problems that arise in finan-

---

3   For the ensuing report, see http://www.iasplus.com/crunch/0804iifbestpractices.pdf.
4   Earlier studies focusing on reputation include Chemmanur & Fulghieri (1994), Smith (1992), Walter & DeLong (1995) and Smith & Walter (1997).

cial intermediation can trigger serious external costs. In recent years the role of various types of financial intermediaries has evolved dramatically. Capital markets and institutional asset managers have taken intermediation share from banks. Insurance activities conducted in the capital markets compete with classic reinsurance functions. Fiduciary activities for institutional and retail clients are conducted by banks, broker-dealers, life insurers and independent fund management companies. Intermediaries in each cohort compete as vigorously with their traditional rivals as with players in other cohorts, competition that has been intensified by deregulation and rapid innovation in financial products and processes. Market developments have periodically overtaken regulatory capabilities intended to promote stability and fairness as well as efficiency and innovation.

It is unsurprising that these conditions would give rise to significant reputational risk exposure for the financial firms involved. For their part, investors in banks and other financial intermediaries are sensitive to the going-concern value of the firms they own, and hence to the governance processes that are supposed to work in their interests. Regulators in turn are sensitive to the safety, soundness and integrity of the financial system, and from time to time will recalibrate the rules of the game. Market discipline, operating through the governance process, interacts with the regulatory process in ways that involve both costs and benefits to market participants and are reflected in the value of their business franchises.

What is reputational risk?

There are substantial difficulties in defining the value of a financial firm's reputation, the extent of damage to that reputation and the origins of that damage, and therefore the sources of reputational risk. Reputation itself may be defined as the opinion (more technically, a social evaluation) of the public toward a person, a group of people, or an organisation. It is an important factor in many fields, such as education, business, online communities or social status. In a business context, it helps drive the excess value of a business firm and such metrics as the market-to-book ratio. However, both precise definition and data are lacking. Arguably many deficiencies in both definition and data can be attributed to the fact that theory development related to corporate reputation has itself been deficient. Such problems notwithstanding, common sense suggests some sources of gain/loss in reputational capital:
–   the starting point – what expectations has the firm set?
–   economic performance – market share, profitability and growth
–   stakeholder interface – shareholders, employees, clients, suppliers
–   legal interface – civil and criminal litigation and enforcement actions.

Consequently, proximate symptoms of sources of loss in reputational capital include:
–   client flight and loss for market share

- investor flight and increase in the cost of capital
- talent flight
- increase in contracting costs.

For practical purposes reputational risk in the financial services sector is therefore associated with the possibility of loss in the going-concern value of the financial intermediary – the risk-adjusted value of expected future earnings. Reputational losses may be reflected in reduced operating revenues as clients and trading counterparties shift to competitors, increased compliance and other costs required to deal with the reputational problem – including opportunity costs – and increased firm-specific risk perceived by the market. Reputational risk is often linked to operational risk, although there are important distinctions between the two. According to Basel II, operational risks are associated with people (internal fraud, clients, products and business practices, employment practices and workplace safety), internal processes and systems, and external events (external fraud, damage or loss of assets, and *force majeure*). Operational risk is specifically not considered to include strategic and business risk, credit risk, market risk or systemic risk, or reputational risk.[5] If reputational risk is bracketed-out of operational risk from a regulatory perspective, then what is it? A possible working definition is as follows:

Reputational risk comprises the risk of loss in the value of a firm's business franchise that extends beyond event-related accounting losses and is reflected in a decline in its share performance metrics. Reputation-related losses reflect reduced expected revenues and/or higher financing and contracting costs. Reputational risk in turn is related to the strategic positioning and execution of the firm, conflicts of interest exploitation, individual professional conduct, compliance and incentive systems, leadership and the prevailing corporate culture. Reputational risk is usually the consequence of management processes rather than discrete events, and therefore requires risk control approaches that differ materially from operational risk.

According to this definition, a reputation-sensitive event might occur which triggers an identifiable monetary decline in the market value of the firm. After subtracting from this market capitalisation loss the present value of direct and allocated costs such as fines and penalties and settlements under civil litigation, the balance can be ascribed to the impact on the firm's reputation. Firms that promote themselves as reputational standard-setters will, accordingly, tend to suffer larger reputational losses than firms that have taken a lower profile. Thus reputational losses associated with identical events according to this definition may be highly idiosyncratic to the individual firm.

---

5 Basel II at http://www.bis.org/publ/bcbs107.htm.

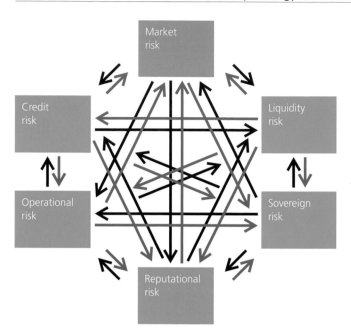

*Fig. 1. A hierarchy
of risks confronting
financial intermediaries*

In terms of the overall hierarchy of risks faced by financial intermediaries, reputational risk is perhaps the most intractable. In terms of figure 1, market risk is usually considered the most tractable, with adequate time-series and cross-sectional data
availability, appropriate metrics to assess volatility and correlations, and the ability
to apply techniques such as Value at Risk (VaR) and risk-adjusted return on capital
(RAROC). Credit risk is arguably less tractable, given that many credits are on the
books of financial intermediaries at historical values. The analysis of credit events in a
portfolio context falls short of market risk, although many types of credits have over
the years become 'marketised' through securitisation structures such as asset backed
securities (ABS) and collateralised loan obligations (CLOs) as well as derivatives such
as credit default swaps (CDS). These are priced in both primary and secondary markets, and transfer some of the granularity and tractability found in market risk to
the credit domain. Liquidity risk, on the other hand, has both pluses and minuses in
terms of tractability – in continuous markets liquidity risk can be calibrated in terms
of bid-offer spreads, although in times of severe market stress and flights to quality,
liquidity can disappear.

If the top three risk domains in figure 1 show a relatively high degree of manage-
ability, the bottom three are less so. Operational risk is a composite of highly manage-
able risks with a robust basis for suitable risk metrics together with risks that represent catastrophes and extreme values – tail events that are difficult to model and in
some cases have never actually been observed. Here management is forced to rely
on either simulations or external data to try to assess the probabilities and poten-

tial losses. Meanwhile, sovereign risk assessment basically involves applied political economy. It relies on imprecise techniques such as stylised facts analysis, so that the track record of even the most sophisticated analytical approaches is not particularly strong – especially under conditions of macro-stress and contagion. As in the case of credit risk, sovereign risk can be calibrated when sovereign foreign-currency bonds and sovereign default swaps (stripped of non-sovereign attributes like external guarantees and collateral) are traded in the market. This leaves reputational risk as perhaps the least tractable of all – with poor data, limited usable metrics, and strong 'fat tail' characteristics.

The other point illustrated in figure 1 relates to the linkages between the various risk-domains. Even the most straightforward of these – such as between market risk and credit risk – are not easy to model or to value, particularly in a bi-directional form. There are 36 such linkages, exhibiting a broad range of tractability. This paper contends that the linkages which relate to reputational risk are among the most difficult to assess and to manage.

Sources of reputational risk

Where does reputational risk in financial intermediation originate? This paper argues that it emanates in large part from the intersection between the financial firm and the competitive environment, on the one hand, and the direct and indirect network of controls and behavioural expectations within which the firm operates on the other, as depicted generically in figure 2.[6] The franchise value of a financial institution as a going concern is calibrated against these two sets of benchmarks. One of them, market performance, tends to be relatively transparent and easy to reward or punish. The other – performance against corporate conduct benchmarks – is far more opaque but potentially critical as a source of risk to shareholders.

Management must work to optimise against both sets of benchmarks. If it strays too far in the direction meeting the demands of social and regulatory controls, it runs the risk of poor performance in the market, punishment by shareholders, and possibly a change in corporate control. If it strays toward unrestrained market performance and sails too close to the wind in terms of questionable market conduct, its behaviour may have disastrous results for the firm, its managers and its shareholders. In the end, striking this balance is a key corporate governance issue.

Such are the rules of the game, and financial intermediaries have to live with them but they are not immutable. There is constant tension between firms and regulators about appropriate constraints on corporate conduct. Sometimes financial intermediaries win battles (and even wars) leading to periods of deregulation. Sometimes it's possible to convince the public that self-regulation and market discipline are powerful

---

6   For an early discussion of external conduct benchmarks, see Galbraith (1973).

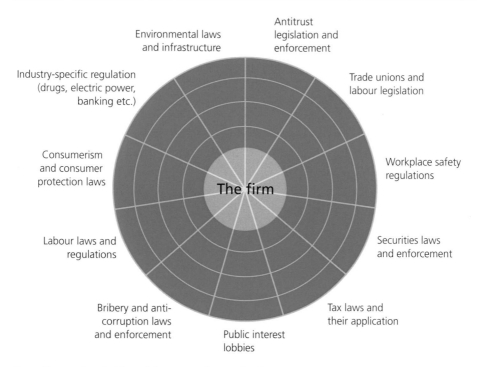

Fig. 2. Reputational risk and the external control web

enough to obviate the need for external control. Sometimes the regulators can be convinced, one way or another, to go easy. Then along comes another major transgression and the constraint system reacts and creates a spate of new regulations. A wide array of interests get into this constant battle to define the rules under which financial business gets done – managers, politicians, the media, activists, investors, lawyers, accountants – and eventually a new equilibrium gets established which will define the rules of engagement for the period ahead.

There are some more fundamental things at work as well. Laws and regulations governing the market conduct of firms are not created in a vacuum. They are rooted in social expectations as to what is appropriate and inappropriate, which in turn are driven by values imbedded in society. These values are rather basic. They deal with lying, cheating and stealing, with trust and honour, with what is right and what is wrong. These are the ultimate benchmarks against which conduct is measured and can be the origins of key reputational losses.

But fundamental values in society may or may not be reflected in people's expectations as to how a firm's conduct is assessed. There may be a good deal of slippage between social values and how these are reflected in the public expectations of business conduct. Such expectations are nevertheless important and the build-up of adverse opinion in the media, the formation of special-interest lobbies and pressure-groups, and the general tide of public opinion with respect to one or another aspect of market conduct can be reputationally debilitating.

Moreover, neither values nor expectations are static in time. Both change. But values seem to change much more gradually than expectations. Indeed, fundamental values such as those noted above are probably as close as one comes to 'constants' in assessing business conduct.

But even in this domain things do change. As society becomes more diverse and mobile, for example, values tend to evolve. They also differ across cultures. And they are sometimes difficult to interpret. Is lying to clients or to trading counterparties wrong? What is the difference between lying and bluffing? Is it only the context that determines how particular behaviour is assessed? The same conduct may be interpreted differently under different circumstances – interpretation that may change significantly over time and differ widely across cultures, giving rise to unique contours of reputational risk.

There is additional slippage between society's expectations and the formation of public policy, and the activities of public interest groups. Things may go on as usual for a while despite occasional media commentary about inappropriate behaviour of a firm or an industry in the marketplace. Then some sort of social tolerance limit is reached. A firm goes too far. A consensus emerges among various groups concerned with the issue. The system reacts through the political process and a new set of constraints on firm develops, possibly anchored in legislation, regulation and bureaucracy. Or the firm is subject to class action litigation.[7] Or its reputation is so seriously compromised that its share price drops sharply.

As managers review the reputational experiences of their competitors, they cannot escape an important message. Most financial firms can endure a credit loss or the cost of an unsuccessful trade or a broken deal, however large, and still survive. These are business risks that firms have learned to detect and limit their exposure toward before the damage becomes serious. Reputational losses may be imposed by external reactions that may appear to professionals as unfocused or ambiguous, even unfair. They may also be a new reading of the rules, a new finding of culpability, something different from the way things were done before. Although regulators and litigants, analysts and the media are accepted by financial professionals as a fact of life, such outsiders can become susceptible to public uproar and political pressure, during which times it is difficult to take the side of an offending financial firm.[8]

In the United States, for example, tighter regulation and closer surveillance, aggressive prosecution and plaintiff litigation, unsympathetic media and juries, and stricter guidelines for penalties and sentencing make it easier to get into trouble and harder to avoid serious penalties. Global brokerage and trading operations, for example, involve hundreds of different, complex and constantly changing products that are

---

7   For a discussion, see Capiello (2006).
8   For a full examination of these issues, see Smith & Walter (1997).

difficult to monitor carefully under the best of circumstances. Doing this in a highly competitive market, where profit margins are under constant challenge and there is considerable temptation to break the rules, is even more challenging. Performance-driven managers, through compensation and promotion practices, have sometimes unwittingly encouraged behaviour that has inflicted major reputational damage on their firms and brought some of them down.

The reality is that the value of financial intermediaries suffers from such uncertain reputation-sensitive conditions. Since maximising the value of the firm is supposed to be the ultimate role of management, its job is to learn how to run the firm so that it optimises the long-term trade-offs between profits and external control. It does no good to plead unfair treatment – the task is for management to learn to live with it, and to make the most of the variables it can control.

The overall process can be depicted in a graphic such as figure 3, representing the firm and its internal governance processes in the centre and various layers of external controls affecting both the firm's conduct and the reputational consequences of misconduct, ranging from 'hard' compliance components near the centre to 'soft' but potentially vital issues of 'appropriate' conduct at the periphery. Clearly, serious reputational losses can impact a financial firm even if it is fully in compliance with regulatory constraints and its actions are entirely legal. The risk of reputational damage incurred in these outer fringes of the web of social control are among the most difficult to assess and to manage. Nor is the constraint system necessarily consistent, with important differences in regulatory regimes (as well as expectations regarding responsible conduct) across markets in which a firm is active, so that conduct, which is considered acceptable in one environment, may give rise to significant reputational risk in another.

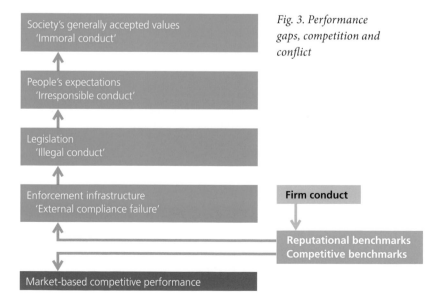

Fig. 3. Performance gaps, competition and conflict

Fig. 4. Reputation-sensitive events in a simple going-concern valuation framework

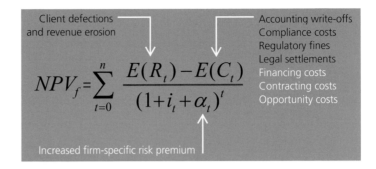

Client defections and revenue erosion

Accounting write-offs
Compliance costs
Regulatory fines
Legal settlements
Financing costs
Contracting costs
Opportunity costs

$$NPV_f = \sum_{t=0}^{n} \frac{E(R_t) - E(C_t)}{(1 + i_t + \alpha_t)^t}$$

Increased firm-specific risk premium

Valuing reputational risk

Recent research has attempted to quantify the impact of reputational risk on share prices during the 1980s and 1990s.[9] Given the nature of the problem, most of the evidence has been anecdotal, although a number of event studies have been undertaken in cases where the reputation-sensitive event was 'clean' in terms of the release of the relevant information to the market.

Figure 4 summarises shareholder value losses in a reputation-sensitive situation involving the aforementioned sources of loss: (1) Client defections and revenue erosion; (2) Increases in monetary costs comprising accounting write-offs associated with the event, increased compliance costs, regulatory fines and legal settlements as well as indirect costs related to loss of reputation such as higher financing costs, contracting costs and opportunity costs – including 'penalty box' suspension by the regulators from particular business activities; and (3) Increases in firm-specific (unsystematic) risk assigned by the market as a result of the reputational event in question. In order to value pure reputational losses, it is necessary to estimate the overall market value loss of the firm to a reputation-sensitive event and subtract from it the monetary losses identified in italics in figure 4.

Consider the following example:[10] On 28 December 1993, the Bank of Spain took control of the country's fourth largest bank, Banco Español de Crédito (Banesto). Subsequently, shares of JPMorgan & Co., a US bank holding company closely involved with Banesto, declined dramatically. Such a reaction appeared inconsistent with market rationality, given that the impact of the event on Morgan's bottom line was trivial (the accounting loss to Morgan was unlikely to exceed $10m after taxes). Perhaps something more than the underlying book value of JPMorgan & Co. was moving the price of the stock, in other words, the central bank takeover of Banesto may have affected the value of Morgan's corporate franchise in some of the firm's core business areas, notably securities underwriting, funds management, client advisory

---

9   For one of the early studies, see Smith (1992).
10   Walter and DeLong (1995).

work and its ability to manage conflicts of interest that can accompany such activities in non-transparent environments.

JPMorgan was involved in Banesto in four ways, in addition to normal interbank transactions relationships:[11]

1   In May 1992, it began raising funds for the Corsair Partnership, L.P., aimed at making non-controlling investments in financial institutions. By February 1993, Morgan had raised over $1 billion from 46 investors that included pension funds and private individuals. Morgan served as General Partner and fund manager, with an investment of $100m. The Corsair Partnership's objective was to identify troubled financial institutions and, by improving their performance, earn a significant return to shareholders in the fund. The Corsair Partnership's first investment, undertaken in February 1993, was a share purchase of $162m in Banesto – giving Morgan a $16.2m equity stake in the Spanish bank.
2   A vice-chairman of JPMorgan served on the Spanish bank's board of directors.
3   Morgan was directly advising Banesto on its financial and business affairs.
4   As part of an effort to recapitalise Banesto, Morgan was lead underwriter during 1993 of two stock offerings that totalled $710m.

Corsair Partnership, L.P. was intended to search for troubled financial institutions in the United States and abroad. The objective was to restructure such institutions by applying Morgan's extensive expertise and contacts. Morgan indicated that Corsair investors could expect a 30% annual return over ten years. Although Morgan had a separate investment banking subsidiary (JPMorgan Securities Inc.), Corsair was believed to be the first equity fund organised and managed by Morgan since the Glass-Steagall Act separated banking and securities activities in 1933 – a separation which eventually lasted until 1999. The business concept of searching for troubled financial institutions emerged from a time of turmoil in the U.S. and foreign banking sectors. When the U.S. banking industry started to improve as a result of a favourable interest rate environment, Corsair ventured abroad. Corsair's first stake in Banesto was taken in February 1993. By August 1993, it had invested $162m (23% of the funds raised) in the Spanish bank. The overall JPMorgan-Banesto relationship is depicted in figure 5.

Banesto's problems stemmed from rapid growth and a convoluted structure of industrial holdings followed by a serious downturn in the Spanish economy. The bank's lending book increased from Pta 0.4 trillion in 1988 to Pta 2.3 trillion in 1991, a period when its competitors were growing at a quarter of that rate. Banesto bid aggressively for deposits, increasing interest rates by 51% while competitors increased theirs by 40%. When the Spanish economy weakened, the bank was stuck with an array of bad loans and the group was further burdened by losses on its industrial holdings.

---

11   For a journalistic account, see *The Wall Street Journal* (1994) and *Euromoney* (1994).

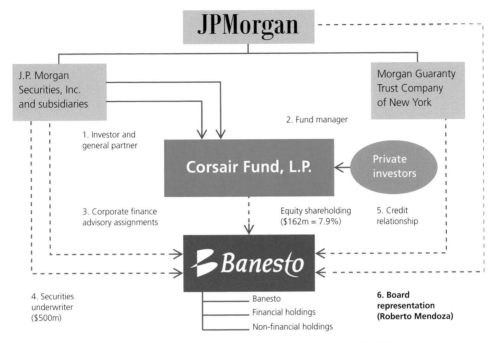

Fig. 5. Reputational risk exposure – JPMorgan and Banco Español de Crédito 1993

In October 1992, a partial audit by the Bank of Spain forced it to lend the troubled institution 'a substantial amount'. An audit released at the end of December 1993 revealed that Banesto assets of Pta 5.5 trillion ($385 billion) were overvalued in excess of Pta 50 billion ($3.5 billion). In April 1994, Banesto was bought for $2.05 billion by Banco Santander, leaving costs of $3.7 billion to be borne by the Spanish banks and by taxpayers.

Morgan had been advising Banesto on various deals since 1987. In July 1992, Morgan's involvement became more intensive when it began advising Banesto on how to raise capital. By August 1993, Morgan had assisted Banesto in two rights issues to raise $710m. During the time of these rights issues, Corsair invested $162m in Banesto. In a letter dated 27 December 1993, Morgan wrote to the Bank of Spain's Governor, outlining how Banesto could continue to raise capital, including a bond issue that Morgan was planning to launch in the first quarter of 1993.

Instead, the Bank of Spain took control of Banesto on the following day, 28 December 1993. Citing mis-management and reckless lending, the Governor of the Bank of Spain justified the action in order to avoid a run on the deposits of the bank, whose share prices were falling sharply on the Madrid Exchange. Given Morgan's multi-faceted involvement in Banesto and potential conflicts imbedded in that relationship, the announcement of the takeover could have had a large effect on the value of Morgan's reputation and business franchise and hence its stock price.

In order to test the impact of the Banesto case on the JPMorgan share price, we use conventional event study methodology.[12] We create a sample prediction of returns on Morgan stock and compare the predicted returns with actual returns on Morgan shares after the Banesto event announcement.[13] The difference is considered the excess return attributable to the event – it represents the difference between what shareholders would have received had they sold their shares in the market 50 days prior to the announcement and the industry's stocks, and what they would have received if they had sold them on subsequent days. If the reputation-effect hypothesis is correct, the market response to the Bank of Spain's announcement on 28 December 1993 should have significantly exceeded the firm's book exposure to Banesto.[14]

Prior to the announcement, Morgan stock behaved as predicted based on its behaviour during the 250 days before the event period. A few days before the announcement, the return began to decline. Thereafter, an essentially steady decline occurred. A cumulative loss of 10% of shareholder equity value was apparent 50 days after the announcement, which translates into a loss in JPM market capitalisation of approximately $1.5 billion versus a maximum direct loss of only $10m from the Banesto failure. This analysis suggests that the loss of an institution's franchise value can far outweigh an accounting loss when its reputation is called into question, a finding similar to that of Smith (1992) in the case of Salomon Brothers, Inc.

Reasons for the adverse market reaction can only be conjectured. The takeover of Banesto could have been seen as compromising Morgan's reputation in precisely those areas key to its future. Inability to turn Banesto around may have called into

---

12   DeLong & Walter (1994). For event study methodology, see Brown and Warner (1985).

13   In order to create this prediction, we regress the daily return of Morgan stock on the daily return on the market index as well as on an industry-group index. The industry group index included 20 financial institutions with characteristics showing some degree of overlap with those of JPMorgan. This is the unweighted average of share prices for Banc One, BankAmerica, Bank of Boston, Bank of New York, Bankers Trust NY, Barnett Bank, Bear Stearns, Chase Manhattan, Chemical Bank, Citicorp, Continental Bank, First Chicago, First Fidelity Bancorp, First Virginia, Merrill Lynch, Morgan Stanley Group, NationsBank, Paine Webber Group, Salomon Inc. and Wells Fargo. We use data from 300 days to 50 days prior to the announcement date (28 December). The resulting coefficients are then multiplied by the returns on the market and industry indices from 50 days prior to 50 days after the announcement, in order to obtain an estimation of the daily stock return during this period. Then, the excess return is calculated at the 'predicted' return minus the actual Morgan stock returns for the period, and the cumulative excess return is plotted. In order to translate these results into the monetary effect on JPMorgan stock, the cumulated excess return is multiplied by the total market value of equity (shares outstanding times price per share) 50 days before the announcement.

14   We regressed Morgan's stock returns against the value-weighted NYSE index and the industry group composed of 20 banking and securities firms. While autocorrelation can be a problem in using daily stock returns, JPMorgan stock was heavily traded, so that daily carryover is unlikely to be significant. Indeed, when we controlled the industry for this potential problem by including the lagged market index as a regression, the resulting coefficient was negative and statistically insignificant. We obtained the following model, estimated over days –300 to –50 prior to the announcement date: $R_{JPMt} = -0.00014 + 0.5766 \cdot R_{Mt} + 0.2714 \cdot R_{Gt} + u_t$ where R_{JPMt} = Return on JPMorgan stock; R_{Mt} = Return on NYSE composite (value-weighted) index; R_{Gt} = Return on group of companies in the same industry. The excess return attributable to the event is the calculated residual (u_t) from 50 days prior to 50 days after the announcement.

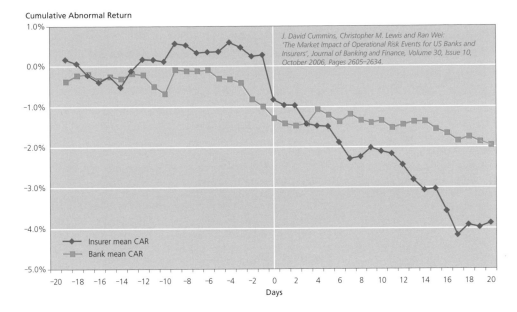

Cumulative Abnormal Return

J. David Cummins, Christopher M. Lewis and Ran Wei:
'The Market Impact of Operational Risk Events for US Banks and
Insurers', Journal of Banking and Finance, Volume 30, Issue 10,
October 2006, Pages 2605–2634.

Fig. 6. Cumulative Abnormal Returns (CARs) for banks and insurers in a large-sample study of operational and reputational events (three-factor models)

question Morgan's ability to successfully advise clients. Banesto as the dominant participation in the Corsair portfolio may have suggested flaws in Morgan's ability to organise and manage certain equity funds. Underwriting stock and placing it with important investor clients raise questions about its ability to judge risks in underwriting securities. Service on Banesto's Board suggests problems with monitoring and the configuration of Morgan's various involvements with Banesto and so also suggests the potential for conflicts of interest or lack of objectivity. Whatever the linkages, here was a case of a financial services firm of exceedingly high standing, which in no way violated any legal or regulatory constraints but whose shares nevertheless appeared to have been adversely affected by market reaction to the way a high-profile piece of business was handled.

In recent years, event studies such as this have yielded a growing body of evidence as to the share price sensitivity to reputational risk. For example, Cummins, Lewis and Wei (2006) undertook a large sample study of operational and reputational events contained in the Fitch OpVar™ database. Figure 6 shows the results in terms of the magnitude of the losses using three-factor estimation models in terms of cumulative abnormal returns (CARs) and number of trading days before and after the announcement. The authors, however, do not distinguish between operational losses and reputational losses, as defined above.

De Fontnouvelle et al. (2006) use loss data from the Fitch OpVar™ and SAS OpRisk™ databases to model operational risk for banks that are internationally active. In a series of robust statistical estimates, they find a high degree of regularity in operational

Event type	SAS OpRisk % of all losses	Percentiles ($m) 50%	75%	95%	Fitch OpVar % of all losses	Percentiles ($m) 50%	75%	95%	Wilcoxon test
All event types	100.0%	6	17	88	100.0%	6	17	93	90.2%
Internal fraud	23.0%	4	10	42	27.0%	6	16	110	1.2%
External fraud	16.5%	5	17	93	16.6%	4	12	70	33.2%
EPWS	3.0%	4	14	–	3.3%	5	11	–	95.0%
CPBP	55.5%	7	20	95	48.1%	7	20	99	96.3%
Damage phys. assets	0.4%	18	–	–	0.3%	20	–	–	92.9%
BDSF	0.2%	36	–	–	0.4%	10	–	–	38.0%
EDPM	1.3%	9	27	–	4.2%	4	11	–	14.6%
Kruskal-Wallis test	6.4E-06				9.1E-05				

Panel B. Losses that occured outside the U.S.

Event type	SAS OpRisk % of all losses	Percentiles ($m) 50%	75%	95%	Fitch OpVar % of all losses	Percentiles ($m) 50%	75%	95%	Wilcoxon test
All event types	100.0%	10	36	221	100.0%	13	46	288	16.7%
Internal fraud	48.5%	9	35	259	42.9%	15	62	381	1.5%
External fraud	15.3%	7	27	–	21.6%	10	28	136	27.3%
EPWS	0.8%	7	–	–	1.6%	2	7	–	75.3%
CPBP	32.6%	14	51	374	28.6%	13	51	359	99.6%
Damage phys. assets	0.0%	–	–	–	0.3%	163	–	–	–
BDSF	0.8%	7	–	–	0.5%	3	–	–	42.3%
EDPM	1.9%	29	–	–	4.6%	5	19	–	8.1%
Kruskal-Wallis test	11.8%				5.5E-05				

Fig. 7. Operational losses by event type[15]

Business line	SAS OpRisk % of all losses	Percentiles ($m) 50%	75%	95%	Fitch OpVar % of all losses	Percentiles ($m) 50%	75%	95%	Wilcoxon test
All business lines	100%	6	17	88	100%	6	17	93	90.2%
Corporate finance	6%	6	23	–	4%	8	23	–	55.8%
Trading and sales	9%	10	44	334	9%	10	27	265	89.2%
Retail banking	38%	5	11	52	39%	5	12	60	73.1%
Commercial banking	21%	7	24	104	16%	8	28	123	13.3%
Payment and settlement	1%	4	11	–	1%	4	11	–	65.8%
Agency services	2%	22	110	–	3%	9	28	–	10.3%
Asset management	5%	8	20	–	6%	8	22	165	80.8%
Retail brokerage	17%	4	12	57	22%	4	13	67	98.0%
Kruskal-Wallis test	2.9E-07				1.0E-12				

Panel B. Losses that occured outside the U.S.

Business line	SAS OpRisk % of all losses	Percentiles ($m) 50%	75%	95%	Fitch OpVar % of all losses	Percentiles ($m) 50%	75%	95%	Wilcoxon test
All business lines	100%	10	36	221	100%	13	46	288	16.7%
Corporate finance	2%	13	–	–	3%	12	27	–	63.3%
Trading and sales	9%	30	125	–	12%	25	66	–	35.3%
Retail banking	41%	6	27	101	44%	9	29	272	10.4%
Commercial banking	30%	15	42	437	21%	35	91	323	2.4%
Payment and settlement	1%	5	–	–	1%	13	–	–	17.7%
Agency services	2%	45	–	–	3%	20	77	–	49.6%
Asset management	3%	5	47	–	5%	7	23	–	90.1%
Retail brokerage	12%	10	42	–	11%	8	34	–	42.6%
Kruskal-Wallis test	2.9E-07				1.0E-12				

Fig. 8. Operational losses by business line[15]

Average MM Abnormal Cumulative Return

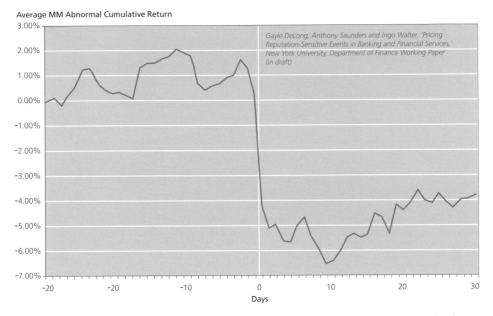

Fig. 9. Reputational Impact and Share Prices. Pilot Study – 49 Events, 1998–2005 (unweighted mean CARs).

losses that are both quantifiable and consistent with large amount of capital increasingly reported by major banks as being held against operational risk – which in many cases exceeds capital held against market risk – figures 7 and 8. The paper also segments the losses by event type and by activity line, as well as by whether or not the operational losses occurred in the United States. The largest losses involved retail and commercial and retail/private banking activities in terms of type of event. As in the case of other studies, the authors do not distinguish the associated accounting losses due to legal settlements, fines, penalties and other explicit operational risk-related costs from reputational losses. As such these estimates are relevant from a regulatory perspective but probably materially understate the losses to shareholders.

In a pilot study of 49 reputation-sensitive events, using the aforementioned definition and excluding operational events, we find negative mean CARs of up to 7% and $3.5 billion, depending on the event windows used. Figure 9 shows the results graphically and the tables in figures 10 and 11 show the numerical results. We do not, however, distinguish between the associated monetary losses and the pure reputational losses.[16]

---

15   Source: De Fontnouvelle, Patrick, Virginia DeJesus-Rueff, John S. Jordan and Eric S. Rosengren (2006). Capital and Risk: New Evidence on Implications of Large Operational Losses. Federal Reserve Bank of Boston. Working Paper. September.

16   Ongoing empirical work on reputation-sensitive financial services events with Gayle DeLong and Anthony Saunders.

Cumulative Abnormal Returns – statistical summary				
Event window	(–5,3)	(–5,10)	(–1,3)	(–1,10)
MEAN	–6,24%	–7,02%	–6,79%	–7,57%
Patell Z-score	–10,02	–7,63	–14,37	–9,41
MEDIAN	–4,59%	–4,92%	–4,55%	–4,96%
Bottom 95% loss	–38,17%	–44,97%	–35,88%	–44,37%
Bottom 99% loss	–62,57%	–47,52%	–63,78%	–48,73%
90% skew	–1,0907	0,1740	–1,2563	0,0538
90% kurtosis	0,0696	–4,6151	0,9144	–4,7431

Fig. 10. Relative CARs – reputational loss pilot study. Source: Gayle De Long, Anthony Saunders and Ingo Walter, 'Pricing Reputation-Sensitive Events in Banking and Financial Services,' New York University, Department of Finance Working Paper (in draft).

Reputational losses in market capitalisation – statistical summary				
Event window	(–5,3)	(–5,10)	(–1,3)	(–1,10)
MEAN	–$3.300.009	–$3.485.131	–$1.765.038	–$1.950.161
p-value	0,0000	0,0013	0,0007	0,0049
MEDIAN	–$984.421	–$555.256	–$700.940	–$616.721
Bottom 95% loss	–$14.875.021	–$24.140.182	–$10.704.029	–$13.227.960
Bottom 99% loss	–$18.375.026	–$28.360.334	–$13.971.351	–$20.261.036
90% skew	–1,5269	0,2562	–0,5088	–1,3309
90% kurtosis	2,4720	–0,4915	0,1960	1,6990

Fig. 11. Absolute CARs – reputational loss pilot study

The only study so far that attempts to identify pure reputational losses is Karpoff, Lee and Martin (2006). The authors attempt to distinguish book losses from reputational losses in the context of US Securities and Exchange Commission enforcement actions related to earnings restatements – 'cooking the books'. The authors review 2,532 regulatory events in connection with all relevant SEC enforcement actions during 1978–2002 and the monetary costs of these actions in the ensuing period, to the end of 2005. These monetary costs are then compared with the cumulative abnormal returns estimated from event studies to separate them from the reputational costs. The results are depicted in figure 12 – note that the reputational losses (66%) are far larger than the cost of fines (3%), class action settlements (6%) and accounting write-offs (25%) resulting from the events in question.

It is likely that the broader the range of a financial intermediary's activities, (1) the greater the likelihood that the firm will encounter exploitable conflicts of interest and reputational risk exposure, (2) the higher will be the potential agency costs facing its clients, and (3) the more difficult and costly will be the safeguards necessary to safeguard the value of the franchise. If this proposition is correct, costs associated with reputational risk mitigation can easily offset the realisation of economies of scope in financial services firms – scope economies that are supposed to generate benefits on the demand side through cross-selling (revenue synergies) and on the supply side

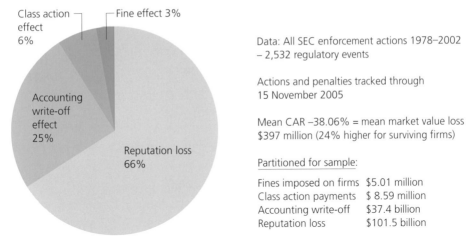

Class action effect 6%

Fine effect 3%

Accounting write-off effect 25%

Reputation loss 66%

Data: All SEC enforcement actions 1978–2002 – 2,532 regulatory events

Actions and penalties tracked through 15 November 2005

Mean CAR –38.06% = mean market value loss $397 million (24% higher for surviving firms)

Partitioned for sample:

Fines imposed on firms	$5.01 million
Class action payments	$ 8.59 million
Accounting write-off	$37.4 billion
Reputation loss	$101.5 billion

Fig. 12. Decomposing CARs related to earnings restatements. Source: Karpoff, Jonathan M., Lee, D. Scott and Martin, Gerald S., 'The Cost to Firms of Cooking the Books' (8 March 2006). Available at http://papers.ssrn.com/sol3/papers.cfm?abstract_id=652121, forthcoming in Journal of Financial and Quantitative analysis.

through more efficient use of the firm's business infrastructure (cost synergies). As a result of conflict exploitation the firm may win and clients may lose in the first instance, but subsequent adverse reputational and regulatory consequences (along with efficiency factors such as the managerial and operational cost of complexity) can be considered diseconomies of scope.

Breadth of engagement with clients may create conflicts of interest that can be multidimensional, and involve a number of different stakeholders at the same time. Several examples came to light during the corporate scandals in the early 2000s. Following the $103 billion bankruptcy of WorldCom in 2002, for example, it appeared that Citigroup – a multifunctional, global financial conglomerate – was serving as equity analyst supplying assessments of WorldCom to institutional and (through the firm's brokers) retail clients, while simultaneously advising WorldCom management on strategic and financial matters. Citigroup's equity analyst at times participated in WorldCom's board meetings. As a major telecommunications-sector commercial and investment banking client, Citigroup maintained an active lending relationship with WorldCom and successfully competed for its securities underwriting business. At the same time, Citigroup served as the exclusive pension fund adviser to WorldCom and executed significant stock option trades for WorldCom executives, while also conducting proprietary trading in WorldCom stock and holding a significant position in the company's stock through its asset management unit. Additionally, Citigroup advised the WorldCom CEO, financed margin purchases of company stock, and provided loans for one of his private businesses.

On the one hand, Citigroup was very successfully engaged in the pursuit of revenue economies of scope (cross-selling), simultaneously targeting both the asset and liability sides of its client's balance sheet, generating advisory fee income, managing assets, and meeting the private banking needs of WorldCom's CEO. On the other hand, that same success caught the firm in simultaneous conflicts of interest relating to retail investors, institutional fund managers, WorldCom executives and shareholders as well as Citigroup's own positions in WorldCom credit exposure and stock trades. WorldCom's bankruptcy triggered a large market capitalisation loss for Citigroup's own shareholders, only part of which can be explained by a $2.65 billion civil settlement the firm reached with investors in May 2004.[17]

It seems plausible that the broader the range of services that a financial firm provides to a given client in the market, and the greater the cross-selling pressure, the greater the potential likelihood that conflicts of interest and reputational risk exposure will be compounded in any given case. When these conflicts of interest are exploited, the more likely they are to damage the market value of the financial firm's business franchise once they come to light. Similarly, the more active a financial intermediary becomes in principal transactions such as affiliated private equity businesses and hedge funds, the more exposed it is likely to be to reputational risk related to conflicts of interest.

Conclusions

This article has attempted to define reputational risk and to outline the sources of such risk facing financial services firms. It then considered the key drivers of reputational risk in the presence of transactions costs and imperfect information and surveyed available empirical research on the impact of reputational losses imposed on financial intermediaries. It concludes that market discipline, through the reputation-effects on the franchise value of financial intermediaries, can be a powerful complement to regulation and civil litigation. Nevertheless, market discipline-based controls remain controversial. Financial firms continue to encounter serious instances of reputation loss due to misconduct despite its effects on the value of their franchises. This suggests material lapses in the governance process.

Dealing with reputational risk can be an expensive business, with compliance systems that are costly to maintain and various types of walls between business units and functions that impose significant opportunity costs due to inefficient use of information within the organisation. Moreover, management of certain kinds of reputational exposure in multifunctional financial firms may be sufficiently difficult to require structural remediation. On the other hand, reputation losses can cause serious damage – as demonstrated by reputation-sensitive 'accidents' that seem to occur repeat-

---

17   Similar issues surfaced in the case of the 2001 Enron bankruptcy. See Batson (2003) and Healy & Palepu (2003).

edly in the financial services industry. Indeed, it can be argued that such issues contribute to market valuations among financial conglomerates that fall below valuations of more specialised financial services businesses. (Laeven & Levine, 2005; Schmid & Walter, 2006).[18] The massive shrinkage of market values of financial firms in 2007–09 depicted in figure 13 certainly embodies reputational damage that will make it even more difficult to recover after the crisis ebbs.

Managements and boards of financial intermediaries must be convinced that a good defence is as important as a good offence in determining sustainable competitive performance. This is something that is extraordinarily difficult to put into practice in a highly competitive environment for both financial services and for the highly skilled professionals that comprise the industry. It seems to require an unusual degree of senior management leadership and commitment. (Smith & Walter, 1997) Internally, there have to be mechanisms that reinforce the loyalty and professional conduct of employees. Externally, there has to be careful and sustained attention to reputation and competition as disciplinary mechanisms. In the end, it is probably leadership more than anything else that separates winners from losers over the long term – the notion that appropriate professional behaviour reinforced by a sense of belonging to a quality franchise constitutes a decisive comparative advantage.

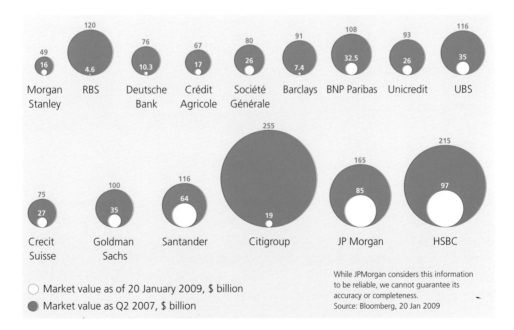

Fig. 13. Declines in market capitalisation of major banks, 2007–09

---

18   See also Kanatas & Qi (2003) and Saunders & Walter (1997).

References

Attorney General of the State of New York. (2003) *Global Settlement: Findings of Fact* (Albany: Office of the State Attorney General).

Batson, Neal. (2003) *Final Report,* Chapter 11, Case No. 01-16034 (AJG), United States Bankruptcy Court, Southern District of New York, July 28.

Brown, Stephen J. and Jerold B. Warner (1985). 'Using Daily Stock Returns: The Case of Event Studies'. *Journal of Financial Economics,* 14, 3–31.

Capiello, S. (2006) 'Public Enforcement and Class Actions Against Conflicts of Interest in Universal Banking – The US Experience Vis-à-vis Recent Italian Initiatives.' Bank of Italy, Law and Economics Research Department, Working Paper.

Chemmanur, Thomas J. and Paolo Fulghieri. (1994) 'Investment Bank Reputation, Information Production, and Financial Intermediation,' *Journal of Finance,* 49, March.

De Fontnouvelle, Patrick, Virginia DeJesus-Rueff, John S. Jordan and Eric S. Rosengren (2006). 'Capital and Risk: New Evidence on Implications of Large Operational Losses.' Federal Reserve Bank of Boston. Working Paper. September.

DeLong, Gayle and Ingo Walter. (1994) 'J.P. Morgan and Banesto: An Event Study.' New York University Salomon Center. Working Paper. April.

Demsky, Joel S. (2003) 'Corporate Conflicts of Interest,' *Journal of Economic Perspectives,* Vol. 17., No. 2, Spring.

Galbraith, John Kenneth. (1973) *Economics and the Public Purpose* (New York: Macmillan).

Herman, Edward S. (1975) *Conflicts of Interest: Commercial Banks and Trust Companies* (New York: Twentieth Century Fund).

Healey, Paul M. and Krishna G. Palepu. (2003) 'The Fall of Enron', *Journal of Economic Perspectives,* Vol. 17., No. 2, Spring.

International Monetary Fund, Global Financial Stability Report (2009). Washington, DC: IMF, April.

Kanatas, George, and Jianping Qi. (2003) 'Integration of Lending and Underwriting: Implications of Scope Economies,' *Journal of Finance,* 58 (3).

Krozner, Randall S. And Philip E. Strahan. (1999) 'Bankers on Boards, Conflicts of Interest, and Lender Liability,' NBER Working Paper No. W7319, August.

Laeven, Luc and Ross Levine. (2005). 'Is There a Diversification Discount in Financial Conglomerates?' C.E.P.R. Discussion Papers No. 5121.

Saunders, Anthony. (1985) 'Conflicts of Interest: An Economic View,' in Ingo Walter (ed.) *Deregulating Wall Street* (New York: John Wiley).

Saunders, Anthony and Ingo Walter. (1997) *Universal Banking In the United States: What Could We Gain? What Could We Lose?* (New York: Oxford University Press).

Schmid, Markus M. and Ingo Walter. (2006) 'Do Financial Conglomerates Create or Destroy Economic Value?' SSRN: http://ssrn.com/abstract=929160.

Schotland, R.A. (1980) *Abuse on Wall Street: Conflicts of Interest in the Securities Markets* (Westport, Ct.: Quantum Books).

Smith, Clifford W. (1992) 'Economics and Ethics: The Case of Salomon Brothers' *Journal of Applied Corporate Finance,* Vol. 5, No. 2, Summer.

Smith, Roy C. and Ingo Walter. (1997) *Street Smarts: Linking Professional Conduct and Shareholder Value in the Securities Industry* (Boston: Harvard Business School Press).

Walter, Ingo. (2004) 'Conflicts of Interest and Market Discipline in Financial Services Firms.' In Claudio Borio, William Curt Hunter, George G. Kaufman, and Kostas Tsatsaronis: *Market Discipline Across Countries and Industries* (Cambridge: MIT Press).

Walter, Ingo and Gayle DeLong, 'The Reputation Effect of International Merchant Banking Accidents: Evidence from Stock Market Data,' New York University Salomon Center, Working Paper, 1995).

About the author

Renée Adams is Professor of Finance at UQ Business School, University of Queensland and Research Associate at the European Corporate Governance Institute (ECGI). Her research focuses on corporate governance, corporate finance and the economics of organisations. She is an expert on corporate boards and the governance of financial institutions. She is the co-author of a forthcoming Journal of Economic Literature survey of the literature on corporate boards with Michael Weisbach and Benjamin Hermalin. She is also the author of a book chapter on bank governance for a handbook on corporate governance to be published by John Wiley & Sons, Inc. Amongst others, she has published in the Journal of Finance, the Journal of Accounting and Economics and the Review of Financial Studies and has a paper forthcoming in the Journal of Financial Economics. She holds a Ph.D. in economics from the University of Chicago and an M.Sc. in mathematics from Stanford University. Contact: r.adams@ business.uq.edu.au

9
Governance and the financial crisis

ABSTRACT Should boards of financial firms be blamed for the financial crisis? When analysing governance, I argue that that it is important to distinguish between banks and non-bank financial firms. I then use a large sample of data on non-financial and financial firms for the period 1996–2007 to analyse board governance in financial firms. I document that the governance of financial firms is, 'on average', not obviously worse than in non-financial firms. Even the issue of executive compensation is not as clear cut as suggested by the media. Although CEOs of financial firms earn on average more than CEOs of non-financial firms, firm-size adjusted CEO pay is lower in banks. I also document that bank directors earned significantly less compensation than their counterparts in non-financial firms and banks receiving bailout money had boards that were more independent than in other banks. I discuss the implications of these findings.

9 Governance and the financial crisis

by Renée Adams

Based on measures of world industrial output, world trade and stock markets, Eichengreen and O'Rourke (2009) argue that the current financial crisis may be worse than the Great Depression on a global scale. Perhaps no one would have been surprised if a crisis of this magnitude originated in an emerging market. Bordo and Eichengreen (2002) provide evidence that most financial crises occur in emerging markets. They describe that there were 139 financial crises between 1973 and 1997, 95 of which occurred in emerging market countries. There are many reasons why investors may lose confidence in emerging markets. If such markets are characterised by weak institutions and poor firm-level governance, then capital outflows and stock market crashes may occur. This is what Johnson, Boone, Breach and Friedman (2000), amongst others, argue happened in the Asian crisis of 1997–1998. But the current financial crisis originated in the USA, a country that is commonly held up as a role model in terms of institutional strength and good governance. For example, the US achieves the highest score on La Porta, Lopez-de-Silanes, Shleifer and Vishny's (1998) anti-director rights index, which measures how well the legal system protects minority shareholders against managers or dominant shareholders.[1] In concurrent work, La Porta et al. (1997) show that common law countries, such as the US, and countries that score higher on their measure of anti-director rights have more developed capital markets. The explanation is that in countries with better protection of shareholders, financiers are willing to invest more money and on better terms for entrepreneurs.

Not only is the US seen as having relatively strong legal institutions, but recent regulation designed to strengthen firm-level governance, the Sarbanes-Oxley Act of 2002 (SOX) and new exchange listing requirements at the NYSE and NASDAQ, have served as models for governance reform around the world. The Sarbanes-Oxley Act and the new listing requirements were a reaction to a series of dramatic corporate and accounting scandals including those at Enron, Tyco and WorldCom. Much of the blame for these scandals was put on boards of directors. For example, in its report on Enron's collapse, the US Senate argued that by not questioning management about the complicated financial transactions Enron was engaging in, the board had failed in its fiduciary duties to shareholders (U.S. House, 2002). Accordingly, both SOX and the NYSE and NASDAQ listing requirements contain a series of provisions designed to strengthen board oversight of management in publicly traded firms.

Currently, for example, a company listed on the NYSE would have to have a major-

---

1   La Porta, Lopez-de-Silanes, Shleifer and Vishny (1998) create this index for 49 countries. The other countries with scores as high as the US are Canada, Chile, Hong Kong, India, Pakistan, South Africa and the UK.

ity of independent directors (new NYSE listing standard), an independent audit committee consisting of at least three members (NYSE) and a financial expert or a reason not to have a financial expert (SOX), a completely independent nominating/corporate governance committee (new NYSE listing standard), a completely independent compensation committee (new NYSE listing standard), regularly scheduled meetings of the non-management directors (new NYSE listing standard) and a yearly meeting of the independent directors (new NYSE listing standard). NASDAQ listed companies are subject to similar requirements, although they do not have to have a separate compensation or nominating committee. In addition, SOX, NYSE and NASDAQ have tightened the definition of independent directors.[2]

Many countries followed the US's suit by tightening governance standards. According to the European Corporate Governance Institute, which maintains a comprehensive database of governance codes around the world (http://www.ecgi.org/codes/index.php). 18 countries published governance codes or made recommendations concerning governance in 2003 alone. Other countries instituted reforms in the ensuing years.

To a certain extent, the fact that the US is considered to have strong investor protection and good governance may help explain why the financial crisis was predicted by so few. But it raises the question whether and to what extent governance can be considered to be a cause of the financial crisis. The recent resignations of several high profile finance executives, for instance Stan O'Neal at Merrill Lynch, Charles Prince at Citigroup and Marcel Ospel at UBS, and the recommendations by several proxy advisors against the re-election of the board at Citigroup, amongst others (see for instance Moyer, 2008) is direct evidence that boards are, at least partly, being blamed for the crisis. The OECD Steering Group on Corporate Governance goes further. It argues that weak governance is a major cause of the financial crisis (Kirkpatrick, 2009). It places much of the blame on board failures in financial firms, in particular,[3] and has launched an action plan to improve corporate governance.[4] Although the UK generally also scores highly on measures of investor protection, bank governance in the UK is also, at least partly, being blamed for the financial crisis. As a result, Sir David Walker has been commissioned to recommend measures to improve board-level governance at banks to the government.[5]

How can it be that governance problems still exist in the US despite strong shareholder protection mechanisms and recent governance reforms? Can boards of financial firms be to blame for the crisis when publicly traded financial firms have to abide by the same governance requirements in SOX and the listing rules as non-financial

---

2   Regardless of exchange listing, all public companies are supposed to abide by SOX which requires that boards are responsible for internal control, audit committees consist entirely of independent directors, audit committees have at least one financial expert, management certifies financial statements and board members face large penalties for corporate accounting fraud.

3   But, it also identifies governance in large, complex non-financial firms as a problem.

4   See http://www.oecd.org/document/48/0,3343,en_2649_34813_42192368_1_1_1_37439,00.html.

5   See http://www.hm-treasury.gov.uk/press_10_09.htm.

firms? This article tries to shed more light on the extent to which the crisis can be attributable to bad financial firm governance, in particular board structure and incentives. It also examines some lessons we can draw for the future as to how financial firm executives should be paid, board competencies, best practices and so forth. Both because the financial crisis originated in the US and because of data limitations this article focuses on publicly traded financial firms in the US. Because financial firms around the world have different activities and structures and face different regulatory constraints, providing an overview of financial firm governance across multiple countries is a complex task that is beyond the scope of this article.[6]

Section 2 discusses what one might expect a well-governed financial firm to look like. Section 3 compares some measures of governance structure across financial and non-financial firms in the US. Section 4 discusses whether bad governance contributed to the crisis. Section 5 discusses lessons for the future.

What is a well-governed financial firm?

Descriptions in the media of what appear to be egregious governance failures at financial institutions have heightened the impression that governance failures play a large part in the financial crisis. For example, boards are being blamed for what appear to be excessive pay packages that executives of financial firms received even while their firms were failing or being bailed out by the government. Morgenson (2009) reports that executives at seven major financial institutions in distress received $464m in performance pay since 2005, while reporting losses of $107 billion since 2007. However, it is important to keep in mind that the media stories often describe individual cases, not the industry as a whole.

To understand the role governance plays in the financial crisis, it is important to get a broader perspective of potential governance problems in the financial industry. Certainly any broad-based policy reform should not be based on the consideration of isolated cases. Ideally, the academic governance literature would provide guidance on the question of whether financial institutions are well governed or not. However, because of the special nature of financial services, most academic papers exclude firms in the financial services from their data and focus on the governance of non-financial firms. Thus, to obtain a picture of the state of governance in the financial service industry, it is useful to directly examine some data on board characteristics and executive compensation. Before turning to the data, however, it is important to have a picture of what the board of a well-governed financial firm should look like.

---

6   Laeven and Levine (2008) provide insights into the governance role of bank ownership structure across
     various countries.

Financial firm governance – the theory

Boards of financial firms have the same legal responsibilities as boards of non-financial firms, that is the duty of care and loyalty. In addition, publicly traded financial firms have to abide by SOX. However, understanding what constitutes an effective governance structure for a financial firm is complicated by several factors. First, as Adams and Mehran (2003) describe, boards of financial firms may face more pressure to satisfy non-shareholder stakeholders than boards of non-financial firms. Regulators, for example, expect boards to act to ensure the safety and soundness of the financial institution, an objective that may not necessarily be in shareholders' best interest. Consistent with the idea that regulators' and owners' interests may diverge, Laeven and Levine (2008) find in a cross-country analysis that the impact of regulation on bank risk-taking depends on a bank's ownership structure. Adams and Mehran (2003, note 6) provide some examples of additional duties regulators impose on bank boards for the purpose of ensuring soundness, which include the adoption of real estate appraisal and evaluation policies (Federal Reserve Board Commercial Bank Examination Manual) and the annual approval of bank risk management policies (Federal Reserve Board Trading Activities Manual).

Second, financial firms are regulated by several different regulators. Investment banks are regulated by the Securities and Exchange Commission (SEC). Thrifts are regulated by the Office of Thrift Supervision. All banks with FDIC-insured deposits are subject to FDIC regulations. Because banks can choose to have a national or state charter and whether or not to be a member of the Federal Reserve, they effectively choose their regulatory authority. National banks are regulated by the OCC, while state banks are regulated either by the Federal Reserve or the FDIC.

The presence of a regulator raises the question of whether regulatory scrutiny complements or substitutes for board-level governance. There is as yet no satisfactory answer to this question. Furthermore, it is not known whether regulators differ in the intensity with which they scrutinise the boards of the firms they examine. Some have argued that regulators may engage in a race to the bottom in order to attract banks with lax restrictions (see for instance Rosen, 2003, 2005 and Whalen, 2002).

Thus, it is possible that some regulators are more lenient in evaluating bank board behaviour than others. For example, even though Federal Reserve Banks in theory penalise directors for poor attendance behaviour, Adams and Ferreira (2008) find that the attendance behaviour of directors of bank holding companies (BHCs) at board meetings is worse than in non-financial firms. Heterogeneity of regulators suggests that board-level governance may not be the same across all types of financial firms and the governance of each type of firm may need to be considered separately.

Third, the financial services industry underwent many recent changes likely to impact board governance. The banking industry underwent an intense period of consolidation in the 1990s through M&A activity. M&A activity affects board structure in several ways. First, it is common to add directors of target firms to the board of the acquirer in friendly acquisitions, as most banking M&As are. Consistent with this, Adams and Mehran (2008) show that M&A activity leads to an increase in BHC board size. Second, M&A activity may have disciplining effects even if it is friendly. Thus, an active market for corporate control may improve board effectiveness. In the 1990s, banks were also increasingly allowed to engage in investment banking activities (culminating with the passage of the Gramm-Leach-Bliley Act in 1999). This created competition for investment banks, which may have put pressure on their boards. Consistent with this idea, Altınkılıç, Hansen and Hrnjić (2007) find that investment banks do not appear to be ineffectively governed during the 1990–2003 period.

These three factors notwithstanding, because their duties to shareholders are the same, the literature generally argues that the same standards should apply to boards of financial firms as to the boards of non-financial firms. The three most commonly studied features of board structure are board independence, board size and the number of other directorships directors hold. The literature generally argues that boards that are more independent, which means they contain more directors without social or business connections to management, should be more effective (see Adams, Hermalin and Weisbach, 2009, for a survey of the board literature). Smaller boards should be more effective because decision-making costs are lower in smaller groups. Because directors may become too busy when they hold more outside directorships, the literature argues that boards are more effective when directors hold fewer outside directorships.

It is less clear from the literature what effective CEO and director compensation should look like. To align their incentives with those of shareholders, CEOs and directors should receive a certain amount of performance-based pay in the form of equity. In addition, holding performance-pay constant, total compensation should increase as risk increases. However, equity incentives may induce managers to take excessive risks. In addition, poorly governed firms may be more likely to overpay their directors. Thus it is not always clear whether a given compensation contract is effective or not. Using an industry study, Philippon and Reshef (2009) argue that bankers were overpaid during the mid-1990s to 2006, however, they do not control for individual firm characteristics, such as size or risk, which may influence compensation packages.

If governance failures at financial firms are partly to blame for the financial crisis, then we might expect that financial firms have worse governance in terms of board characteristics and incentives than non-financial firms.

In a sample of data on 35 BHCs ending in 1999, Adams and Mehran (2003) find that BHCs have larger boards, more independent directors and lower performance-based pay for CEOs than non-financial firms. Thus, in their sample, banks could be considered better governed than non-financial firms in some aspects such as independence, but worse in other aspects such as board size and performance-pay. Since their sample predates the Gramm-Leach Bliley Act of 1999 which may have influenced governance structures in the industry, it is worthwhile making a similar comparison in a larger sample of more recent data, as I do in the next section.

The governance of financial firms – what are the facts?

Data

To compare board characteristics and incentives in financial firms and non-financial firms, I use the Riskmetrics director database from 1996 to 2007.[7] This is an unbalanced panel of director-level data for Standard & Poor's (S&P) 500, S&P MidCaps, and S&P SmallCap firms. It contains information on directors from company proxy statements or annual reports, such as whether the director is classified as independent and the number of other directorships each director holds. I merge this data to Compustat to obtain financial information and SIC codes, which I use to identify financial firms and banks. I obtain data on CEO compensation and director compensation from ExecuComp. All compensation numbers are measured in thousands and are adjusted to 2007 dollars using the CPI-U. Using this data, I construct a dataset containing 18,542 firm-year level observations of which 14.57% (2,702 observations) correspond to financial firms. Of the financial firm observations, 42.39% (1,136 observations) belong to banks. All of the firms in this database are publicly traded firms and all banks are BHCs, thus this data does not allow me to shed any light on the governance of private financial firms. The number of BHCs represented in the sample varies over time from a low of 86 to a high of 105. The number of non-bank financial firms varies from 106 to 158. Although the number of banks in the sample is small relative to the banking industry, the banks in this sample represent a large fraction of industry assets. For example, according to the FDIC, in 2007 there were 7,282 FDIC-insured commercial banks in the US with a total of 11,176 billion in assets. In 2007, there are 93 banks in my sample with assets comprising 69.15% of total commercial bank assets (7,728,449m). Thus, although the comparisons I make and conclusions I draw need not apply to all banks, they are relevant for understanding potential governance failures at banks which are likely to matter the most for the crisis.

Comparing banks to non-financial firms

Because of the reasons outlined in the previous section, I compare the governance characteristics of banks and non-bank financials to those of non-financials separately. Table 1 provides comparisons for banks and non-financial firms. Table 2 provides the same comparisons for non-bank financial firms. Columns I-III of panel A, table 1 show comparisons of means of board independence, board size and the average number of

---

7   I use this data only after extensive cleaning.

Entire sample 1996–2007						
	Board independence	Board size	# directorships	Total CEO compensation	Fraction equity-based pay CEO	Director compensation
	I	II	III	IV	V	VI
Panel A: Simple comparison of means for banks versus non-banks						
Bank	0.045***	4.624***	-0.306***	313.872***	0.001	-45.347***
	[8.07]	[56.86]	[14.73]	[6.33]	[0.11]	[3.56]
Constant	0.652***	9.145***	0.838***	1,181.748***	0.525***	176.188***
	[455.20]	[434.75]	[156.92]	[94.50]	[217.00]	[50.78]
Obs.	16976	16976	14268	14503	14408	1669
R-sq.	0.004	0.16	0.015	0.003	0	0.008
Panel B: Comparison of means for banks versus non-banks after controlling for firm size as proxied by the natural logarithm of assets						
Bank	-0.012**	2.310***	-0.775***	-760.329***	-0.131***	-106.091***
	[2.07]	[30.26]	[38.52]	[15.81]	[13.21]	[8.31]
Firm size	0.023***	0.935***	0.193***	428.237***	0.053***	28.763***
	[25.21]	[79.56]	[62.34]	[59.05]	[35.18]	[14.22]
Constant	0.481***	2.267***	-0.590***	-2,010.768***	0.132***	-52.474***
	[69.17]	[25.65]	[25.21]	[36.41]	[11.54]	[3.20]
Obs.	16616	16616	13956	14503	14408	1669

Table 1. Comparison of selected governance characteristics of banks to non-financial firms
The table shows a comparison of selected governance characteristics between banks and non-financial firms. The sample consists of an unbalanced panel of 18,542 firm-year level observations for S&P 1500 firms for the period 1996–2007 in the Riskmetrics 'Director Database'. Data on assets and SIC codes is from Compustat. Data on CEO and director compensation is from ExecuComp. SIC codes were used to classify firms into banks, financial firms and non-financial firms. 14.57% (2,702 observations) correspond to financial firms. Of the financial firm observations, 42.39% (1,136 observations) belong to banks. Riskmetrics classifies directors as independent if they have no business relationship with the firm, are not related or interlocked with management and are not current or former employees. 'Board independence' is the number of independent directors on the board divided by board size. 'Number of directorships' is the average number of directorships in other for-profit companies per director. 'Total CEO compensation' is the sum of 'Salary', 'Bonus', 'Other Annual', 'Total Value of Restricted Stock Granted', 'Total Value of Stock Options Granted' (using Black-Scholes), 'Long-Term Incentive Pay-outs', and 'All Other Total'. Fraction equity-based pay is 1-(Salary+Bonus)/Total CEO Compensation. 'Director compensation' is only available for the years 2006 and 2007 and is the average compensation paid to directors. All compensation numbers are measured in thousands and have been converted to 2007 dollars using the CPI-U. 'Firm size' is the natural logarithm of the book value of assets. Assets are measured in millions. 'Bank' is a dummy variable equal to 1 if the sample firm is a bank. Observations vary because of missing data. Absolute values of t-statistics are in brackets. Asterisks indicate significance at 0.01 (***), 0.05 (**), and 0.10 (*) levels.

2007 data only						
	Board independence	Board size	# directorships	Total CEO compensation	Fraction equity-based pay CEO	Director compensation
	VII	VIII	IX	X	XI	XII
Panel A: Simple comparison of means for banks versus non-banks						
Bank	-0.007	3.055***	-0.471***	-124.31	-0.076**	-62.866***
	[0.56]	[12.82]	[8.42]	[0.44]	[2.16]	[2.81]
Constant	0.773***	9.019***	0.914***	1,189.315***	0.715***	193.931***
	[233.18]	[138.34]	[59.73]	[16.27]	[79.32]	[33.77]
Obs.	1269	1269	1269	702	702	700
R-sq.	0	0.115	0.053	0	0.007	0.011
Panel B: Comparison of means for banks versus non-banks after controlling for firm size as proxied by the natural logarithm of assets						
Bank	-0.034***	1.369***	-0.777***	-932.094***	-0.204***	-124.593***
	[2.74]	[6.66]	[14.67]	[3.25]	[5.97]	[5.52]
Firm size	0.014***	0.858***	0.154***	361.605***	0.057***	27.586***
	[6.77]	[25.27]	[17.67]	[8.67]	[11.53]	[8.40]
Constant	0.665***	2.383***	-0.277***	-1,750.618***	0.250***	-30.25
	[40.58]	[8.90]	[4.02]	[5.06]	[6.08]	[1.11]
Obs.	1261	1261	1261	702	702	700

directorships per director for banks and non-financial firms. Columns IV–VI show comparisons for total CEO compensation, including the value of equity-based pay, the fraction of equity-based pay in total CEO compensation, a measure of performance pay and average director compensation.[8] Columns I–VI are for the entire sample from 1996 to 2007. Columns VII–XII repeat the comparison for 2007 data only. Observations vary because of missing data. The coefficient on the Constant term measures the sample mean for non-financial firms. The coefficient on 'Bank' ('Non-bank financial firm') measures the difference between banks (non-bank financial firms) and non-financial firms. Asterisks indicate statistical significance, with ***, ** and * indicating significance at the 1%, 5% and 10% level, respectively.

From columns I–VI of table 1, panel A, we see that banks have on average more independent boards, larger boards, fewer outside directorships, higher total CEO compensation and lower director compensation than non-financial firms. While board size is larger in banks, Adams and Mehran (2008) do not find that larger bank board size has detrimental effects on shareholder value. Thus, on average, banks do not appear to be worse governed than non-financial firms, except possibly in terms of compensation. However, it is well known that wages increase in firm size and banks are generally much larger than non-financial firms in terms of assets. Thus, in panel B, I perform the same comparisons after controlling for firm size, as proxied by the natural logarithm of the book value of assets.[9] The picture looks slightly different now.

---

8   The director compensation measures are only available for 2006 and 2007 due to changes in reporting requirements.

Entire sample 1996–2007					
Board independence	Board size	# directorships	Total CEO compensation	Fraction equity-based pay CEO	Director compensation
I	II	III	IV	V	VI

Panel A: Simple comparison of means for non-bank financial firms versus non-financial firms

	Board independence	Board size	# directorships	Total CEO compensation	Fraction equity-based pay CEO	Director compensation
Non-bank financial firm	-0.025***	1.214***	-0.031*	914.753***	0.030***	6.091
	[5.21]	[18.01]	[1.75]	[18.99]	[3.46]	[0.61]
Constant	0.652***	9.145***	0.838***	1,181.748***	0.525***	176.188***
	[448.45]	[452.40]	[157.00]	[86.25]	[215.10]	[51.74]
Obs.	17406	17406	14667	14775	14679	1750
R-sq.	0.002	0.018	0	0.024	0.001	0

Panel B: Comparison of means for non-bank financial firms versus non-financial firms after controlling for firm size as proxied by the natural logarithm of assets

	Board independence	Board size	# directorships	Total CEO compensation	Fraction equity-based pay CEO	Director compensation
Non-bank financial firm	-0.061***	-0.243***	-0.310***	127.647***	-0.060***	-32.330***
	[12.17]	[4.06]	[18.55]	[2.81]	[7.00]	[3.33]
Firm size	0.024***	0.908***	0.179***	447.951***	0.051***	28.199***
	[26.71]	[85.74]	[60.17]	[58.66]	[35.36]	[15.01]
Constant	0.477***	2.462***	-0.490***	-2,157.734***	0.143***	-47.992***
	[70.85]	[30.83]	[21.70]	[37.04]	[12.91]	[3.14]
Obs.	17041	17041	14348	14775	14679	1750
R-sq.	0.041	0.314	0.202	0.208	0.079	0.114

Table 2. Comparison of selected governance characteristics of non-bank financial firms to non-financial firms

The table shows a comparison of selected governance characteristics between non-bank financial firms and non-financial firms. The sample is described in table 1. Riskmetrics classifies directors as independent if they have no business relationship with the firm, are not related or interlocked with management and are not current or former employees. Board independence is the number of independent directors on the board divided by board size. 'Number of directorships' is the average number of directorships in other for-profit companies per director. 'Total CEO compensation' is the sum of 'Salary', 'Bonus', 'Other Annual', 'Total Value of Restricted Stock Granted', 'Total Value of Stock Options Granted' (using Black-Scholes), 'Long-Term Incentive Payouts', and 'All Other Total'. Fraction equity-based pay is 1-(Salary+Bonus)/Total CEO Compensation. 'Director compensation' is only available for the years 2006 and 2007 and is the average compensation paid to directors. All compensation numbers are measured in thousands and have been converted to 2007 dollars using the CPI-U. 'Firm size' is the natural logarithm of the book value of assets. Assets are measured in millions. 'Non-bank financial firm' is a dummy variable equal to 1 if the sample firm is a financial firm that is not a bank. Observations vary because of missing data. Absolute values of t-statistics are in brackets. Asterisks indicate significance at 0.01 (***), 0.05 (**), and 0.10 (*) levels.

2007 data only						
	Board independence	Board size	# directorships	Total CEO compensation	Fraction equity-based pay CEO	Director compensation
	VII	VIII	IX	X	XI	XII

Panel A: Simple comparison of means for non-bank financial firms versus non-financial firms

Non-bank financial firm	-0.032***	0.950***	-0.082*	345.224	-0.038	4.903
	[3.32]	[4.97]	[1.84]	[1.49]	[1.47]	[0.31]
Constant	0.773***	9.019***	0.914***	1,189.315***	0.715***	193.931***
	[230.67]	[137.16]	[59.26]	[14.47]	[77.93]	[34.17]
Obs.	1332	1332	1332	751	751	749
R-sq.	0.008	0.018	0.003	0.003	0.003	0

Panel B: Comparison of means for non-bank financial firms versus non-financial firms after controlling for firm size as proxied by the natural logarithm of assets

Non-bank financial firm	-0.054***	-0.156	-0.273***	-199.976	-0.105***	-29.755*
	[5.49]	[0.96]	[6.46]	[0.89]	[4.23]	[1.90]
Firm size	0.016***	0.838***	0.142***	421.555***	0.052***	26.722***
	[8.26]	[25.79]	[16.82]	[9.73]	[10.84]	[8.85]
Constant	0.647***	2.535***	-0.185***	-2,238.021***	0.294***	-23.227
	[41.42]	[9.86]	[2.76]	[6.20]	[7.41]	[0.92]
Obs.	1322	1322	1322	751	751	749
R-sq.	0.057	0.347	0.179	0.115	0.138	0.095

Controlling for size, bank boards are less independent, larger, have fewer outside directorships, have lower total CEO compensation, lower incentive pay for the CEO and lower director compensation than non-financial firms. These results hold even in the 2007 data. Clearly, firm size is an important factor influencing governance characteristics, consistent with findings in Boone, Fields, Karpoff and Raheja (2007), Coles, Daniel and Naveen (2008), Lehn, Patro and Zhao (2008) and Linck, Netter and Yang (2008). For example, the coefficient on 'Bank' in the board size comparison decreases by roughly 50% in Column II after controlling for firm size, and the compensation numbers decrease significantly. Although bank boards look less independent after controlling for firm size, the difference in magnitude (1.2% in Column I) is small compared to mean independence of 65.2% for non-financial firms. Based on these comparisons, it would be difficult to argue that banks are clearly poorly governed. In particular, much of the media attention focuses on 'excessive' total pay and performance pay in financial firms, yet, on average, bank CEOs earn less and have less performance pay than CEOs of non-financial firms of similar size.

---

9   These results are obtained by OLS regressions of the dependent variables on a dummy variable which is defined to be one for banks and 0 for non-financial firms and the proxy for firm size.

Comparing financial firms to non-financial firms

The picture looks slightly different when we compare non-bank financial firms to non-financial firms in table 2. Comparing panel A (simple means) to panel B (means after controlling for firm size), it is clear that firm size again has a significant effect on governance characteristics. After controlling for firm size, non-bank financial firms have less independent boards, smaller boards, fewer outside directorships, lower performance pay and lower director compensation than non-financial firms. However, now CEO compensation is still higher than in non-financial firms even after controlling for size, although the difference is not significant in 2007. Based on these comparisons, one might argue that board independence is too low and CEO pay too high in financial firms. However, it is not clear whether better performance on other governance measures such as smaller boards and fewer outside directorships could compensate for these differences.

Some conclusions one can draw from the discussion in this section are as follows: First, banks and non-bank financial firms do not have exactly the same governance characteristics. Second, although executive pay in financial firms appears larger than in non-financial firms, much of this difference is driven by differences in firm size. Of course, CEOs at some financial firms may have received what many consider to be unfair or excessive pay. However, on average, financial CEO pay does not seem to be excessive. If anything, the phenomenon of 'excess pay' for CEOs appears to be driven by non-bank financial firms. Third, although financial firms score worse on some governance characteristics than non-financial firms, they score better on others. Thus, it would be difficult to say that, on average, financial firms are clearly governed worse than non-financial firms. Nevertheless, it is possible that these differences are partly to blame for the problems facing financial firms. I turn to this issue in the next section. A final and potentially very important observation is that director pay is significantly lower in banks, even without controlling for firm size. This fact is also pointed out in Adams and Ferreira (2008). From table 1, panel A, column XII, one can see that in 2007 average bank director compensation is lower than average non-financial director compensation by $62,866.00. Since average director compensation in non-financials is $193,931, bank director compensation is roughly 32% lower. This may have serious consequences for governance. If the pool of director candidates is the same for financial firms and non-financial firms, then the better qualified candidates may prefer to join the boards of firms which pay more. Of course, factors other than pay play a role in an individual's decision to join a board, but this finding raises the possibility that the pool of bank directors may be worse along some dimensions than the pool of non-financial firm directors.

Is governance to blame for the problems at banks?

In the previous section, I showed that the governance of financial firms differs from that of non-financial firms. Although it is not clear whether these differences are in the direction of worse governance, the differences are almost all statistically significant. Naturally, one would like to know whether these differences contributed to the problems facing financial firms. In general, this is not an easy question to answer, especially not now while the crisis is still playing itself out. One way of trying to answer this question is to compare the governance characteristics of financial firms which received bailout money from the Federal Government to those that did not. The Troubled Assets Relief Program (TARP) is designed to strengthen the financial system by enabling the government to purchase or insure up to $700 billion in troubled assets. Although the US treasury does not describe it as a bailout program (see http://www.financialstability.gov/roadtostability/capitalpurchaseprogram.html), many observers refer to it as such (see for instance the TARP entry on Wikipedia). Nevertheless, it is clear that recipients of TARP money face some sort of financial difficulties, thus a comparison of characteristics of firms with and without TARP assistance may still be illustrative. In addition, I can eliminate firms that failed or were acquired from my sample to ensure I am comparing institutions which are essentially healthy that did not receive bailout funds to those that did.[10]

As of 10 April 2009, only six non-bank financial firms received TARP assistance. Thus, I limit the comparison to banks. To determine which of the banks in my sample received TARP funds, I matched their names and locations to the names and locations in the list of TARP recipients maintained by the New York Times (http://projects.nytimes.com/creditcrisis/recipients/table). Of the 93 banks in my sample in 2007, 56 received bailout funds in either 2008 or beginning of 2009 (until 10 April 2009). To determine whether any of the remaining institutions failed or were acquired, I first merge my sample to data from Osiris gathered on 8 April 2009. This dataset contains information about the current status of institutions. For the 16 institutions whose status I was unable to verify using Osiris, I checked institutional histories on the National Information Center's website,[11] a website containing information on banks collected by the Federal Reserve System. Four institutions in my sample disappeared by 2009. Two were acquired (Wachovia and National City Corp), one closed (Downey Financial Corp) and one failed (Franklin Bank Corp). I eliminate these observations from my sample. In table 3, I perform a similar comparison for 2007 as in tables 1 and 2 after restricting my sample to banks. The coefficients on 'bailout' measure differences in governance characteristics for banks that received TARP money in 2008 and 2009 to those that did not. Because the sample is so small, I do not also report the results af-

---

10   Unfortunately, it is not possible to determine whether any of the banks in my sample applied for TARP money but were turned down.

11   http://www.ffiec.gov/nicpubweb/nicweb/NicHome.aspx.

2007 data only						
	Board independence	Board size	# directorships	Total CEO compensation	Fraction equity-based pay CEO	Director compensation
	VII	VIII	IX	X	XI	XII
Bailout	0.037*	1.340**	0.265***	302.53	0.163*	-3.699
	[1.68]	[2.13]	[2.66]	[0.78]	[1.97]	[0.14]
Constant	0.745***	11.303***	0.246***	861.947**	0.517***	131.295***
	[42.67]	[22.64]	[3.11]	[2.69]	[7.59]	[5.98]
Obs.	89	89	89	44	44	44
R-sq.	0.031	0.049	0.075	0.014	0.085	0

Table 3. Comparison of selected governance characteristics for sample banks receiving bailout money to surviving sample banks which did not

The table shows a comparison of selected governance characteristics in 2007 between sample banks that received bailout money from the US government in 2008 and beginning of 2009 (up until 10 April 2009) and sample banks which survived until April 2009 and did not receive bailout money. I define a bank to have received bailout money if it received funds from the US government under the Troubled Asset Relief Program (TARP). The underlying sample is described in table 1. To determine which of the banks in my sample received TARP funds, I matched their names and locations to the names and locations in the list of TARP recipients maintained by the New York Times (http://projects.nytimes.com/creditcrisis/recipients/table). Of the 93 banks in my sample in 2007, 56 received bailout funds in either 2008 or beginning of 2009 (until 10 April 2009). To determine whether any of the remaining institutions failed or were acquired, I first merge my sample to data from Osiris gathered on 8 April 2009. This dataset contains information about the current status of institutions as of beginning of 2009. For the 16 institutions whose status I was unable to verify using Osiris, I checked institutional histories on the National Information Center's website, a website containing information on banks collected by the Federal Reserve System. Four institutions in my sample disappeared by 2009. Two were acquired, one closed and one failed. I eliminate these observations from my sample. 'Board independence' is the number of independent directors on the board divided by board size. 'Number of directorships' is the average number of directorships in other for-profit companies per director. 'Total CEO compensation' is the sum of 'Salary', 'Bonus', 'Other Annual', 'Total Value of Restricted Stock Granted', 'Total Value of Stock Options Granted' (using Black-Scholes), 'Long-Term Incentive Payouts', and 'All Other Total'. Fraction equity-based pay is 1-(Salary+Bonus)/Total CEO Compensation. 'Director compensation' is the average compensation paid to directors. 'Bailout' is a dummy variable equal to 1 if the sample firm received bailout money from the US government. Observations vary because of missing data. Absolute values of t-statistics are in brackets. Asterisks indicate significance at 0.01 (***), 0.05 (**), and 0.10 (*) levels.

ter controlling for firm size. The coefficients have similar signs, although they are less significant, except for director compensation which becomes significantly negative.

Table 3 indicates that banks with TARP funds have more independent boards, larger boards, more outside directorships and greater incentive pay for CEOs. Some of these results are consistent with the idea that TARP banks have worse governance. In particular, the fact that TARP banks had higher performance-pay for CEOs is consistent with the idea that performance pay may have led executives of banks to take on too much risk. The coefficient on the number of directorships is also consistent with potentially worse governance since taking on too many directorships can lead directors to become too unfocused. The fact that TARP banks have larger boards is also a potential indication of worse governance. Perhaps the most surprising result, however, is the finding that TARP banks have boards which are more independent.

What is going on here? There are several possible explanations. For example, it is possible that the governance of TARP banks is not worse than that of non-TARP banks. However, given that a large part of the problems at banks was caused by securitisation, a different explanation seems more plausible. An independent director, by definition, is a director who has not worked for the bank and has no business dealings with the bank. Because of potential conflicts of interest, independent directors are generally not employees of other financial firms. What this means is that independent directors are less likely to have an in-depth knowledge of the internal workings of the banks on whose boards they sit. They are also less likely to have the financial expertise to understand the complexity of the securitisation processes banks were engaging in or to assess the associated risks banks were taking on. Thus, although board independence is generally seen to be a good thing, in the case of banks, greater independence may be a bad thing because a more independent board will not have sufficient expertise to monitor the actions of the CEO. This finding is also consistent with Guerrera and Larsen (2008) who describe that more than two-thirds of the directors at eight large US financial institutions did not have any significant recent experience in the banking industry and more than half had no financial service experience at all.

If we accept that independence may be a sign of poor governance, then the conclusion to be drawn from table 3 is that it appears that TARP-banks were indeed governed worse than non-TARP banks. However, because the message from tables 1 and 2 is mixed, which means that banks and financial firms do not necessarily appear to be worse governed than non-financial firms, it is not obvious what the policy implications are. I discuss some lessons we can learn in the next section.

What lessons can we learn?

Because it is the duty of the board to oversee management and many bank and financial firm managers led their banks to the brink of failure, boards of financial firms clearly share some responsibility for the crisis. But the question is how much of the blame should they shoulder? Popular opinion, fuelled by stories of what seems to be excessive executive compensation at financial firms, might argue: a lot. However, it is important to keep several facts in mind, particularly when considering potential policy changes.

First, few people predicted the financial crisis. Rushe (2008) describes, for example, how little attention was paid to Nouriel Roubini's prediction that problems in the sub-prime mortgage market would trigger a financial crisis. While the boards of financial firms should have better information than outsiders, directors are generally not experts on the economy. Thus, it seems unreasonable to expect that they should have been better able to predict the problems financial firms would face than academics, regulators and financial analysts. Rodrik (2009) argues in his blog that blame should be spread more widely than bankers to the broader economics and policy-making community.

Second, financial firms are regulated. It is still not clear whether regulators substitute or complement board-level governance. However, it is possible that directors perceive a substitution effect. It is not hard to imagine that a bank director does not understand all risk implications of particular transactions but agrees to them anyhow, because he assumes that regulators would identify any potential problems.

Third, the data in this paper show that governance in financial firms is, on average, not obviously worse than in non-financial firms. While financial firm governance may appear worse in some dimensions, it appears better in others. *Ex post,* it is easy to argue that governance problems occurred, but *ex ante* it is not clear that boards of financial firms were doing anything much different from boards in other firms. Even the issue of executive compensation is not as clear cut as it appears at first. CEOs of financial firms do earn significantly more than CEOs of non-financial firms, but this is no longer true for bank CEOs once firm size is accounted for. However, two particular findings suggest that banking firm governance may have been worse, but in ways that might not have been predicted *ex ante*. Banks receiving bailout money had boards that were more independent and bank directors earned significantly less compensation than their counterparts in non-financial firms. What this suggests is that board independence may not necessarily be beneficial for banks. Independent directors may not always have the expertise necessary to oversee complex banking firms. The fact that bank directors earn so much less than their counterparties in non-financial firms raises the possibility that the pool of bank directors is different from the pool of directors of non-financial firms. Further research is needed to examine why director pay is so low in banking.[12] However, regardless of whether low pay is a sign of a governance problem or not, it seems clear that if it is not increased it will be difficult to attract candidates to bank director positions given the additional duties expected of them in the future.

While representing a large fraction of banking industry assets, the sample in this paper is relatively small compared to the number of institutions in the financial sector. Furthermore, many more factors affect governance characteristics than firm size. Thus, much more research is needed to understand the extent to which governance contributed to the financial crisis.

---

12   It is possible that regulators do not look favourably on pay raises for directors.

Nevertheless, the simple analysis in this paper is still suggestive that board-level governance of publicly traded financial firms may have played a part in the crisis. However, it also suggests that the recent governance reform movement embodied in SOX and the NYSE and NASDAQ listing standards may be as much to blame. SOX and the listing standards place a lot of emphasis on director independence. However, it is not clear that director independence is always beneficial because independent directors lack information (see also the arguments in Adams and Ferreira, 2007). The problem may be exacerbated for financial firms because of the complex nature of their businesses. This is a point the OECD Steering Committee on Corporate Governance (Kirkpatrick, 2009) also makes. It argues that independence at financial firms may have been overemphasised at the expense of qualifications. Guerrera and Larsen (2009) also discuss the fact that SOX made it more difficult for financial firms to hire suitable directors with financial expertise because of the perception of conflicts of interest.

Some tentative policy implications can be drawn as follows: By placing too much emphasis on independence, SOX and recent listing standards may have worsened board governance at publicly traded financial firms. Not only are financial firms complex, but the financial industry is complex due to the existence of different regulators. Until the governance of financial firms is better understood, it may be better not to impose restrictions on the governance of financial firms. Further regulating governance may have unintended negative side effects, as SOX may have had. In addition, compliance with regulation may be costly. For example, some evidence exists that suggesting complying with the requirements in SOX and the listing standards, imposed significant costs on small firms. Amongst others, Leuz, Triantis and Wang (2008) find that compliance with SOX was a significant factor in companies' decisions to delist in 2002–2004. Since the majority of firms in the banking sector are relatively small, at this point in time, not having an explicit governance policy for financial firms may be better than imposing additional restrictions on governance.

However, it may be beneficial for financial firms to try to increase the financial sophistication of their boards, either through hiring new directors or through additional education. Because financial firms, such as Citigroup, are already trying to change their boards, it is not clear that increasing the financial sophistication of the board requires regulation. Regardless of their expertise, it is also likely that directors of financial firms will not need any regulatory prodding to ask more tough questions of management in the future. One problem that remains is the issue of director compensation. It may be difficult to institute pay raises for directors of financial firms given the current outrage over executive compensation at AIG and other large financial firms. But, to ensure that financial firms retain good directors and attract good candidates, directors of financial firms should be adequately compensated for the difficulties of their duties and the additional costs they bear in undertaking any additional training.

References

Adams, R. and D. Ferreira, 'A Theory of Friendly Boards', *Journal of Finance,* 62(1), pp. 217–250, 2007.

Adams, R. and D. Ferreira, 'Does Regulatory Pressure Provide Sufficient Incentives for Bank Directors? Evidence from Directors' Attendance Records,' University of Queensland Working Paper, 2008.

Adams, R., B. Hermalin and M. Weisbach, 'The Role of Boards of Directors in Corporate Governance: A Conceptual Framework and Survey', forthcoming *Journal of Economic Literature,* 2009.

Adams, R. and H. Mehran, 'Corporate Performance, Board Structure and its Determinants in the Banking Industry', University of Queensland Working Paper, 2008.

Adams, R. and H. Mehran, 'Is Corporate Governance Different for Bank Holding Companies?' *Economic Policy Review,* 9(1), pp. 123–142, 2003.

Altınkılıç, O., R. Hansen and E. Hrnjić, 'Investment Bank Governance', Tulane University Working paper, 2007.

Boone, A., L. Field, J. Karpoff and C. Raheja, 'The determinants of corporate board size and composition: An empirical analysis', *Journal of Financial Economics* 85, pp. 66–101, 2007.

Bordo, M. and B. Eichengreen, 'Crises now and then: What lessons from the last era of financial globalization?' NBER Working Paper No. 8716, 2002.

Coles, J., N. Daniel, L. Naveen, 'Boards: Does one size fit all?' *Journal of Financial Economics* 87, pp. 329–356, 2008.

Eichengreen, B. and K. O'Rourke, 'A Tale of Two Depressions', http://www.voxeu.org/index.php?q=node/3421, 6 April 2009.

Guerrera, F. and P. Larsen, 'Gone by the board? Why bank directors did not spot credit risks', *Financial Times,* 25 June 2008.

Johnson, S., P. Boone, A. Breach and E. Friedman, 'Corporate Governance in the Asian Financial Crisis', *Journal of Financial Economics,* 58, pp. 141–186, 2000.

Kirkpatrick, G., 'The Corporate Governance Lessons from the Financial Crisis,' *Financial Markets Trends,* 2009/1, OECD, 2009.

La Porta, R., F. Lopez-de-Silanes, A. Shleifer R. Vishny, 'Legal Determinants of External Finance', *Journal of Finance* 52(3), pp.1131–1150, 1997.

La Porta, R., F. Lopez-de-Silanes, A. Shleifer R. Vishny, 'Law and Finance', *Journal of Political Economy* 106(6), pp. 1113–1155, 1998.

Laeven, L. and R. Levine, 'Corporate Governance, Regulation, and Bank Risk Taking', forthcoming *Journal of Financial Economics,* 2008.

Lehn, K., S. Patro and M. Zhao, 'Determinants of the Size and Composition of U.S. Corporate Boards: 1935–2000', forthcoming *Financial Management,* 2008.

Leuz, C., A. Triantis and T. Wang, 'Why Do Firms Go Dark? Causes and Economic Consequences of Voluntary SEC Deregistrations', *Journal of Accounting and Economics* 45, pp. 181–208, 2008.

Linck, J., J. Netter and T. Yang, 'The determinants of board structure', *Journal of Financial Economics* 87(2), pp. 308–328, 2008.

Morgenson, G., 'After Huge Losses, a Move to Reclaim Executives' Pay', The New York Times, p. BU1, 22 February 2009.

Moyer, L., 'Off with their heads', Forbes.com, available at http://www.forbes.com/2008/04/11/shareholders-banking-iss-biz-wallcx_lm_0411proxy.html, 4 April 2008.

Philippon, T. and A. Reshef, 'Wages and Human Capital in the U.S. Financial Industry: 1909–2006', NBER Working Paper No. 14644, 2009.

Rodrik, D., 'Simon Johnson's morality tale', http://rodrik.typepad.com/dani_rodriks_weblog/2009/03/simon-johnsons-morality-tale.html, 30 March 2009.

Rosen, R., 'Is three a crowd? Competition among regulators in banking', *Journal of Money, Credit, and Banking* 35, pp. 967–998, 2003.

Rosen, R., 'Switching primary federal regulators: Is it beneficial for U.S. banks?' *Economic Perspectives* 29, 2005.

Rushe, D., 'Nouriel Roubini: I fear the worst is yet to come', *The Sunday Times*, 26 October 2008.

U.S. House, Committee on Governmental Affairs, Permanent Subcommittee on Investigations, 107th Cong., 2d sess., 'The role of the board of directors in Enron's collapse', Committee Print, 2002.

Whalen, G., 'Charter Flips by National Banks', Office of the Comptroller of the Currency, E&PA Working Paper 2002–1, 2002.

About the author

 *Kjell Johan Nordgard was born in Oslo, Norway in 1960. He holds a master's degree in electronics and computer science from The Norwegian University of Science and Technology and a bachelor degree in finance from Copenhagen Business School. Kjell Nordgard has worked most of his professional life servicing the financial industry in different capacities spanning software design and development, through to consultancy and training, and management and strategy. For the last 15 years he has worked for SimCorp, a global provider of software solutions to the financial sector, where he is Senior Vice President with overall responsibility for global market support activities such as product strategy, product management, partnerships, professional services and global marketing. Prior to this, Kjell Nordgard headed up SimCorp's operations in the UK and Ireland as Managing Director of SimCorp Ltd.*

Contact: kjell.j.nordgard@simcorp.com

10
Governance, risk and compliance in the financial sector: an IT perspective in light of the recent financial crisis

ABSTRACT Following the recent financial crisis questions are being raised as to how this could be possible in a sector that is already so heavily regulated. With reference to some examples as to what triggered the crisis, the article highlights the increasing complexity in the financial sector. As the financial industry is heavily reliant on information, sound corporate governance should ensure that this information is available at all times and at all required levels in the organisation. IT governance must therefore be an integral part of corporate governance, and ensure the company implements an IT infrastructure that supports and extends the company's business model. This article discusses some main guidelines for such an IT infrastructure, with particular focus on support for risk management and compliance controls.

10 Governance, risk and compliance in the financial sector: an IT perspective in light of the recent financial crisis

by Kjell Johan Nordgard

Background

Corporate governance as a term has probably been around since the inception of corporate structures as we know them today where a company is organised with a board of directors, executive directors, managers and employees and where capital is raised through external investors in an open capital marketplace.

Throughout the centuries, the term has gained renewed interest whenever there was an economic crisis of one sort or another. Scholars, legislators and business professionals have discussed the reasons for the crisis and introduced different initiatives through legislation and best practice with the aim of reducing the likelihood of ending up in a similar situation again.

At the beginning of the new millennium, scandals at corporate giants like Enron, Xerox and WorldCom cost investors billions of dollars and sent shockwaves through the US economy. The situation became even worse when it became clear that the culprits were not limited to local boards, directors or key employees of the collapsed companies, but were also to be found amongst representatives of the external entities supposed to control these companies. Arthur Anderson, the Big 5 auditing firm, was, in the aftermath of the Enron collapse, found guilty of criminal charges in respect of their auditing services and was subsequently wound down as a company.

The situation thus sparked a rising mistrust to the commercial system in general. Clearly, there was a need to firm up on everything from corporate roles and responsibilities through to better control functions and more transparent reporting to reveal the true state of affairs. In short, there was a need to enforce a better culture for corporate governance. To address this issue, the US Senate passed the Sarbanes-Oxley Act on 30 July 2002. Sarbanes-Oxley, commonly referred to as SOX, sets new standards for all public US companies in order to ensure sound corporate governance and hence protect investors from losses incurred by bad management and controls, hidden risk exposure or outright fraud.

So did the scholars, legislators and business professionals succeed in plugging the holes this time? Clearly, the SOX initiative was necessary and serves as a good common denominator for the elements that need to be in place in order to implement sound corporate governance in a firm. However, despite this initiative, the world found itself in turmoil again in 2008. This time however, the scenario was different. Whereas the situation that led to the SOX legislation came on the back of a burst

'IT bubble' in 2000 followed by revelations of cheating, fraud and bad management amongst leading so-called blue chip companies, this time the crisis struck at the very foundation of the capitalist system, the institutions in charge of ensuring availability of capital and funding of corporate activity.

The crisis started amongst the financial institutions with sudden collapses of big international banks like Bear Sterns, Lehman Brothers and Merrill Lynch as well as leading insurance groups like AIG and US mortgage lenders Fannie Mae and Freddie Mac. It spread like fire across the entire international capital marketplace and the financial system was brought to the brink of collapse. Liquidity was frozen by banks out of fear of losing out on counterparty default, and as the mechanisms for funding commercial activity had essentially come to a standstill, the entire corporate world was soon embroiled in a deep and spiralling crisis.

This article is not about analysing the reasons for the recent financial crisis. The point to make here is, however, that yet again we have experienced a major international crisis sparked by lack of sound corporate governance, this time amongst the financial institutions. These institutions are regulated by several laws and directives: Basel II, Solvency II, UCITS III+IV, IFRS to mention just a few of them. As an example, Basel II is aimed at the banking sector which by its nature is based on a leveraged business model. Client deposits are geared up and passed on as loans to other clients, and profit is made from the margin between borrowing and lending interest rates. The shareholder capital is normally quite slim, must be at least 8% according to Basel I, and hence the gearing is high. Basel II sets out rules to control such gearing levels in that it requires banks to set aside capital to provide cover for financial and operational risks. When calculating the market risk of assets, the different investments have different weights according to their risk profile. So far so good, but then the creativity started in the banking sector. Through a combination of reckless use of securitisation and outright circumvention of legislation, banks managed to increase their leverage far beyond the limits set out by Basel II.[1]

Securitisation is in the essence a very sensible construction, its purpose being to transfer risk from those responsible for raising funds over to the vast universe of private investors. In the case of the current financial crisis, securitisation was achieved by packaging loans as collateral for complex structured instruments like collateralised debt obligations (CDOs), collateralised loan obligations (CLOs) and collateralised mortgage obligations (CMOs). The collateral for these vehicles was corporate loans and mortgage loans, many of them so-called sub-prime loans associated with high yield and high risk. Through a number of intermediaries, who all charged lucrative fees for their contribution to the party, these loans ended up on the shelves of the investment banks as AAA rated investment products in the shape of CDOs. Now, how the rating bureaus managed to rate these products at this high level is subject

---

1 Acharya, Viral & Richardson, Matthew (eds.). 2009. 'Restoring Financial Stability'.

for a study on its own and will not be discussed in this article. An important point to bear in mind is though that most of the underlying mortgage loans had so called 'teaser rates' attached to them. This means that the borrowers received a grace period at the beginning with very low interest and redemption rates. The interest rates then increased over the life of the loans to provide a sufficient average spread commensurate with the risk associated with the collateral. This structure may work well during periods of rising house prices, but when the markets turn and the mortgage payers over time are hit by the higher interest rates and are confronted with the redemption payments, the result is the shattering of the very foundation of structures like CDOs.

As CDOs were AAA rated, they were, in Basel II terms, cheaper for the banks to keep on their balance sheets compared to the corresponding underlying loans. Hence the CDOs represented potential for further gearing. Furthermore, because they gave a decent return, and probably also because top management in the investment banks didn't really understand the true exposure they undertook to the housing market through these structures, almost 50% of these issues remained on the balance sheets of the banks rather than being sold off to the market. This left the US financial sector with an effective total exposure to the housing market of roughly $5.8 trillion before the bubble burst. The problem was that securitisation stopped half way through the process; the packaging succeeded but the distribution failed. Moreover, in order to circumvent regulation aimed at capping leverage, banks created so-called special purpose vehicles (SPVs)[2] to which they offloaded some of the riskier assets they would otherwise hold on their books. In order for these SPVs to remain creditworthy, their structures involved liquidity and credit enhancements from the banks which effectively kept the risk with the banks but took the assets off of their balance sheets. The consequences of this excessive gearing and overexposure to the housing market is today, with hindsight, painfully clear. The economy turned, people could no longer pay their mortgages, the CDOs collapsed and so did many of the main players in the financial marketplace. The financial system itself only survived due to massive government interventions worldwide.

One conclusion to draw from all of this is probably just that history repeats itself and that in between we have to live with the consequences of some of the less appealing characteristics of human nature, greed and selfishness. As long as there are rules and legislation in place to ensure proper behaviour, there will always be creative brains searching for loopholes and ways to avoid these rules in order to maximise personal short-term winnings. Another conclusion is however that the world is becoming more and more complex, and it is getting increasingly difficult for boards of directors and executive managements of international firms to fully understand the true risks their companies are exposed to. This complexity is particularly dominant in the financial sector. The innovation in the financial markets is beyond anything ex-

---

2   The most commonly used SPVs were asset backed commercial paper (ABCP) conduits and structured investment vehicles (SIV).

perienced in other commercial segments, with new products being launched more or less on a daily basis. Furthermore, as the players in the financial markets are so tightly interlinked, the default of a single or a few large players soon becomes the problem of the whole sector. This is what is referred to as 'systemic risk,' and this is effectively what has become a reality in the financial marketplace during the past couple of years primarily due to lack of control on CDOs and other structured financial products – this lack of control again due to lack of proper corporate governance in the financial sector.

The increased complexity in international business calls for new initiatives in order to ensure proper conduct and transparency as to how companies are managed and how risks and results are reported. The drive towards better corporate governance continues and legislation and directives need to take the ever-changing marketplace into consideration in order come up with effective measures. It is however increasingly clear that no board or CEO, CIO, COO or other manager in a sizeable international firm can come even close to execution of good corporate governance without the right IT support to provide sufficient insight into the true exposures and risk of the enterprise at all times. Conversely, the one and only reason for investing in IT must be to support the business values of the corporation and create a competitive advantage. This is why the terms corporate governance and IT governance must go hand in hand – you cannot properly do one without the other.

Corporate governance, risk and compliance

Corporate governance is a broad term, but basically covers the responsibility of boards of directors and senior executive management to create organisational policies and structures to ensure transparency as to how business is conducted, what results the business produces and what risks the business undertakes in order to produce these results. A proper governance policy ensures strategies to implement systems to monitor and record current business activity, takes steps to ensure compliance with legislation and agreed policies, and provides for corrective action in cases where the rules have been ignored or misconstrued.

Business author Gabrielle O'Donovan makes the following definition in her book 'The Corporate Culture Handbook':[3]

"Corporate governance is an internal system encompassing policies, processes and people, which serves the needs of shareholders and other stakeholders, by directing and controlling management activities with good business savvy, objectivity, accountability and integrity. Sound corporate governance is reliant on external marketplace commitment and legislation, plus a healthy board culture which safeguards policies and processes."

---

3   O'Donovan, Gabrielle. 2007. 'The Corporate Culture Handbook'.

Traditionally, the subject of corporate governance has been tightly linked to shareholder value. And even though the more recent initiatives and debates have widened the focus more to other stakeholders internal and external to the firm, one may say that shareholder value is still the ultimate goal. At the end of the day, corporate governance is about improving the way business is conducted and hence increasing the value for the owners of the company, that is to say the shareholders. Furthermore, as share prices are determined by expected future performance of a firm, an important aspect of corporate governance is to ensure long-term value creation through a stable and robust business model rather than always going for the quick wins.[4]

The Organisation for Economic Co-operation and Development (OECD) has issued a paper outlining some overall principles to guide sound corporate governance:[5]

1 **Ensuring the basis for an effective corporate governance framework**
    The corporate governance framework should promote transparent and efficient markets, be consistent with the rule of law and clearly articulate the division of responsibilities among different supervisory, regulatory and enforcement authorities.
2 **The rights of shareholders and key ownership functions**
    The corporate governance framework should protect and facilitate the exercise of shareholders' rights.
3 **The equitable treatment of shareholders**
    The corporate governance framework should ensure the equitable treatment of all shareholders, including minority and foreign shareholders. All shareholders should have the opportunity to obtain effective redress for violation of their rights.
4 **The role of stakeholders in corporate governance**
    The corporate governance framework should recognise the rights of stakeholders established by law or through mutual agreements and encourage active co-operation between corporations and stakeholders in creating wealth, jobs, and the sustainability of financially sound enterprises.
5 **Disclosure and transparency**
    The corporate governance framework should ensure that timely and accurate disclosure is made on all material matters regarding the corporation, including the financial situation, performance, ownership, and governance of the company.
6 **The responsibility of the board**
    The corporate governance framework should ensure the strategic guidance of the company, the effective monitoring of management by the board, and the board's accountability to the company and the shareholders.

---

4   Moorman, Theodore Clark. 2005. 'Corporate governance and long-term stock returns'.
    http://repositories.tdl.org/tdl/handle/1969.1/2341?show=full.
5   'OECD Principles of Corporate Governance', 2004.

These six headings cover a lot of ground and involve the corporation dealing with internal as well as external factors and stakeholders. Internal stakeholders are members of the board, executive directors, managers and employees of the firm. Examples of external stakeholders are legislators, shareholders, potential investors, market analysts, creditors, suppliers and customers. In reality, external stakeholders cover anybody with an interest in the firm in one way or another.

Whereas internal factors can be controlled and managed by the firm, the external ones are different in that they can normally only be observed and not controlled. The external factors will of course influence the internal factors in that the corporate governance policy must make sure internal measures are taken to address external factors and to live up to expectations from internal and external stakeholders. It's about creating trust and transparency for all stakeholders, and this trust is worth money. A study from McKinsey & Company shows that 78% of investors are willing to pay a surplus premium for well-governed companies.[6] The study also reveals that 56% of the population rates corporate governance either as more important than or on a par with financial results such as profit and performance of a company when it comes valuing its shares.

Examples of some internal and external factors and are given in table 1.

Internal	External
Internal reporting requirements	Competition
Measures to avoid conflict of interest between board, directors and managers	Debt covenants
Internal authorisation policy	Demand for financial statements
Organisational structures	Government regulations
Internal control procedures	Labour market
Risk strategies	Media pressure
Risk and performance reporting	Takeovers
Remuneration policy and incentive schemes	Market economy
IT infrastructure	

Table 1. Internal and external factors to be covered by corporate governance

One of the reasons corporate governance is so high up on the investors' radar is that sound governance strategies and policies make up the very foundation for proper management of risk and compliance in a company. Risk management is the process by which an organisation sets the risk tolerance, identifies potential risks and prioritises the tolerance for risk based on the organisation's business objectives. Risk management leverages internal controls to manage and mitigate risk throughout the organisation. Compliance is the process that records and monitors the controls (be they physical, logical or organisational) needed to enable compliance with legislative directives, industry mandates or internal policies.

---

6   'McKinsey Global Investor Opinion Survey on Corporate Governance'. 2002.

This relationship has given rise to the acronym 'GRC' – governance, risk management and compliance – a buzz term often used when analysing companies. The connection is the following: if proper governance is not in place, meaningful risk management and compliance controls cannot be achieved. In the same vein, if targeted risk management policies, tools and metrics are not in place – there is no foundation for controlling a company's compliance with internal and external legislation, directives and policies.

The main focus in this article will be on the GRC relationship in financial institutions, as this is where we can find the roots of many of the issues that led to the current market crisis.

GRC in financial institutions

At the end of the day, corporate governance is essentially about creating trust. Without trust, the markets become unstable and unpredictable. This is particularly true in the financial markets. Professor Paul Verdin from Leuven University illustrates this with the trust/confidence hypothesis as depicted in figure 1.

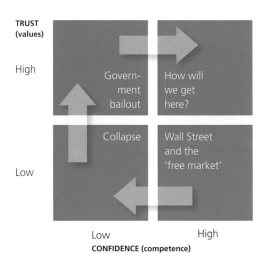

Fig. 1. The Trust-Confidence hypothesis, Paul Verdin, Ph.D., Department of Economics, Strategy and Innovation, University of Leuven.

Confidence is about having faith that people employed in a firm have the right competencies for conducting their business. Trust is about having the faith that these people also have sound values which will ensure business is conducted in an ethical and trustworthy manner. The hypothesis is that people, up to the recent crisis, had confidence that directors and employees in the financial sector knew their markets well and had the competencies to run their business professionally. The trust in the industry's general morale and values has however traditionally been at a low level. The scenario therefore starts in the lower right-hand corner.

With the collapse of major financial institutions in 2008, confidence in the segment's professional competence faded, and the segment moved on to the left. The ensuing government bailouts and rescue packages were followed up by tighter scrutiny and direct control of the financial firms, and this increased the trust in the sense that people would now feel the entities in the financial industry no longer would be able to cut corners and chase risky short-term gains to the extent they had been used to. This is where we are today. Some trust has been induced, but the confidence in the industry's competence to run their business is still missing.

A quick look at figure 2 supports this hypothesis. The graph shows the development of the PHLX KBW Bank Sector Index, a capitalisation-weighted index composed of 24 geographically diverse stocks representing banks and leading financial institutions. When some of the big US banks started to crack in September 2008, the whole financial sector rapidly plummeted back to levels not seen since 1996.

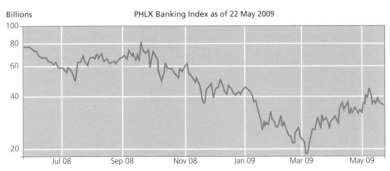

Fig. 2. 'PHLX KBW Bank Sector Index, Sep 2008–May 2009'.
Source: http://finance.yahoo.com.

Even though the dollar-losses revealed by the announced bankruptcies in September/October 2008 were shocking, they cannot on their own justify this total collapse across an entire segment. The sharp decline was triggered mainly by lost confidence. When the highly educated, highly paid money managers at the big prestigious Wall Street banks could make such blunders and lose that amount of money, then surely that must be a common denominator of the whole sector? The graph shows a slight pickup from March 2009 when the different government aid and rescue packages started to work and some level of trust was established. But the challenge now is for the financial industry to move to the upper right-hand corner of figure 1. Confidence in the industry's professionalism must be re-gained and the trust level must not falter in the process. The only way to make this move is to improve corporate governance in the sector.

So what makes the challenges faced by the financial sector so different from those of companies in other sectors? One important factor is that many financial institutions, such as banks and hedge funds, are highly geared and can swiftly change their risk profile. Such changes in the risk profile can, through actions undertaken by portfolio managers and traders, easily take place without transparency to top management. Another aspect of the financial marketplace is that the innovation in the sector is unique with new complex products coming to market in a continuous stream. The complexity of these instruments makes it extremely difficult to fully assess true risks and exposures. Finally, the actors in the financial marketplace are heavily interlinked and dependent on each other. The attention level and information flow is enormous and the international marketplace for trading of financial products operates around the clock every day of the year. All these factors give rise to extra challenges for this sector in respect of good corporate governance, particularly in the areas of authorisation and control or compliance, accurate risk management and transparent reporting of true performance and exposures.

These areas are of course highly interrelated. In financial business, there is a direct relationship between risk and performance. In portfolio theory, risk is the price of the return on the investments. The art of portfolio management therefore consists of balancing an acceptable risk profile against the target returns on the funds under management. In a broader perspective, one may say the same applies to corporate governance; the objective is to maximise a company's long-term performance in a controlled risk environment. Ernst & Young puts risk, compliance and performance in a corporate governance perspective in their Risk and Control Framework Model as depicted in figure 3.

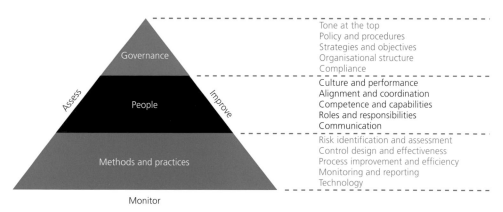

Fig. 3. The Risk and Control Framework Model. 2008. Ernst & Young.

Governance is in this model driven by the 'tone at the top', in other words the tone set by directors and managers to uphold ethics, integrity and responsibility in the organisation. Employees and managers at all levels pay close attention to behaviour and priorities of their bosses, and will therefore do what they witness their bosses doing. The governance policy must take its outset in the 'tone at the top' and drive high-level strategies, initiatives and structures to be put in place in order to ensure that people practice a risk and performance conscious working culture based on the right competence and capabilities and facilitated by defined tools in the shape of proven methods and practices. The model is dynamic, in that all levels in the pyramid must continuously be subjected to monitoring, assessment and improvement. The world is a dynamic playing field, especially when it comes to financial markets, and any model for governance must therefore be adaptable in order to remain effective when external factors change (see table 1).

The model illustrates the tight relationship between overall corporate strategies and directives, corporate working culture and practical tools. Corporate governance is about striking a balance and assuring coherence between these three levels in the pyramid. Furthermore, the model illustrates that good methods and practices represent the very foundation of effective execution of governance policies, and that these tools must be developed to deliver against the overall business objectives of the firm. It is in this area, the support of methods and practices, that the IT architecture of a company really comes to the fore and can mean the difference between a successful or a failed corporate governance practice.

Based on the arguments as presented in this section, the main topics for discussion in this article will be on challenges in respect of sound corporate governance of financial institutions in the areas of risk management, and compliance including internal and regulatory reporting. These will be the main drivers in rebuilding confidence in the sector and starting the move towards the upper right-hand corner in figure 1. Furthermore, the aspects of corporate governance discussed in this article will primarily be those where IT can make a significant difference. Therefore, even though topics like morale, ethics, remuneration policy and organisational measures to avoid conflicts of interest are highly relevant themes for any discussion around corporate governance, they will not be discussed further here.

First, however, we will briefly visit the main elements covered by the term IT governance.

IT governance

"A good craftsman is recognised by the quality of his tools"'
IT governance is about ensuring that the organisation has the IT tools and processes in place to implement sound corporate governance. The definition of IT governance from the IT Governance Institute (ITGI)[7] reads;

"IT governance is the responsibility of the board of directors and executive manage-ment. It is an integral part of enterprise governance and consists of the leadership and organisational structures and processes that ensure that the organisation's IT sustains and extends the organisation's strategies and objectives."

The ITGI works on best practice models for IT strategies in corporations, and their framework builds on the following five principles in respect of sound IT governance:
- strategic alignment of IT with the business
- value delivery of IT
- management of IT risks
- IT resource management
- performance measurement of IT.

Fig. 4. The five pillars of IT governance[8]

These five pillars of good IT governance are depicted in figure 4. The ITGI stresses that IT governance is an integral part of corporate governance and is, like any other governance matter, ultimately the responsibility of board members and executives in a corporation. IT governance and IT strategies have traditionally been too low on the agenda for boards and executive management, but it has become increasingly clear that no modern organisation above a certain level of complexity can implement sound and effective corporate governance without also dealing professionally with IT governance.

The five pillars of the ITGI model can be described as follows:

Strategic alignment
The implemented IT governance framework needs to ensure an IT strategy and infra-

---

7   ITGI is a not-for-profit organisation set up in 1998 to assist enterprise leaders in their responsibility to make IT successful in supporting the enterprise's mission and goals.
8   'Board Briefing on IT Governance, 2. edition'. 2003. ITGI.

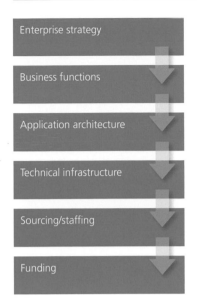

Fig. 5. The process of establishing a business-driven IT infrastructure[9]

structure that supports and extends the enterprise's business objectives. This framework furthermore needs to ensure that the IT governance principles are pervasive in the organisation and that roles, responsibilities and reporting lines are clearly defined. IT strategy and IT operations need to be aligned with the business strategy and the business operations. This is essential in order for the IT function to deliver recognisable value to the business, and it means that roles and responsibilities of IT departments need to ensure a close link to the appropriate business objectives at all levels in the organisation. The principle that business strategies must drive the IT strategies is depicted in figure 5.

Executive management needs quality data with the right level of detail in order to manage the business and secure sound corporate governance. It is down to the IT framework, through the different levels in the organisation, to ensure these data are delivered as required.

Value delivery

This pillar emphasises the importance of tangible business value delivered by the IT function. A prerequisite for such value generation is, as explained, that IT strategy is closely aligned with business strategy. It is therefore important the business strategy sets out specific success criteria and that such success criteria, to the extent possible, are measurable and expressed as key performance indicators (KPIs). This will create

---

9   'Board Briefing on IT Governance, 2. edition'. 2003. ITGI.

visibility all the way up to board level as to how the IT function succeeds in delivering against the business objectives of the firm. As examples of assessment areas for value delivery, ITGI mentions:

- fit for purpose, meeting with business requirements
- flexibility to adopt to future requirements
- throughput and response times
- ease of use, resiliency and security
- integrity, accuracy and currency of information
- time-to-market
- cost and time management
- partnering success
- skill set of IT staff.

With this, examples of relevant KPIs for a financial institution within these assessment areas could be:
- maximum implementation shortfall[10] for:
  - investment strategies such as switch strategies, alpha strategies etc.
  - cash injections in portfolios
  - new investment products such as funds etc.
  - rebalancing of portfolios against models or benchmarks
  - financial risk assessments
  - performance attribution reporting
  - new accounting principles
  - new client services such as analysis, reporting and advice.
- minimum acceptable straight through processing (STP) rates;[11]
- maximum allowed failing rate for post-trade processing;
- maximum allowed time for transfer of back office data to middle and front office which could include coupons, redemptions, corporate actions among other things;
- maximum allowed time to aggregate data in order to produce full exposure reports towards any company and counterparty;
- maximum allowed down time. International firms will normally require 24x7 availability;
- maximum cost of operation measured in basis points of assets under management.

---

10 Implementation shortfall is here defined as the time it takes from when a decision is made until the result of the decision is fully recognised on the books, that is structures set up, trading is done, settlement is done and profit and loss calculated.

11 STP rates can be defined as automation level. The extent to which the IT infrastructure supports the principle of only entering the same data once. STP rate of 100% then means that all relevant data are only entered once, and when entered they apply for the whole operation across front, middle and back office.

The business values generated by a successfully implemented IT strategy will be instrumental in giving a financial institution the competitive edge in a tough and challenging marketplace. The process through which the IT function delivers true business value is depicted in figure 6. An effective IT infrastructure will maximise flexibility and minimise implementation time, defined as the time from strategic business decisions being taken to when financial business impact of these decisions can be observed.

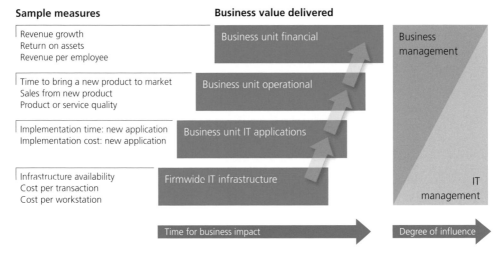

Fig. 6. The value impact of IT infrastructure

Risk management

Risk management is a broad topic and will be covered in more detail in following sections. Regulators are specifically concerned about operational and systemic risk, within which technology risk and information security issues are prominent. In the context of the ITGI framework, the focus is on management of operational risks in respect of the IT infrastructure. In this category, the following issues are important for most businesses, but particularly important for banks and other financial institutions:

- availability of IT infrastructure including disaster recovery;
- availability of user applications;
- connectivity to counterparty applications such as static data, trading, settlement, reconciliation and so forth;
- access to on-line market information such as prices, interest rates, foreign exchange (FX) rates and market news;
- availability of data across the enterprise;
- quality, consistency and timeliness of data across the enterprise;
- confidentiality and restriction of data access to authorised people.

Effective risk management begins with a clear understanding of the enterprise's appetite for risk, including the high-level risk exposures of the enterprise. This focuses all risk management effort and, in an IT context, impacts future investments in technology, the extent to which IT assets are protected and the level of assurance required. Having defined risk appetite and identified risk exposure, strategies for managing risk can be set and responsibilities clarified. Dependent on the type of risk and its significance to the business, management and the board may choose to:

– mitigate – implement controls, such as acquiring and deploying security technology to protect the IT infrastructure;
– transfer – share risk with partners, outsource or transfer to insurance coverage;
– accept – formally acknowledge that the risk exists and monitor it.

As a minimum, risk should at least be analysed, because even if no immediate action is taken, the awareness of risk will influence strategic decisions for the better. Often, the most damaging IT risks are those that are not well understood.

Resource management
This category is about efficiency in providing the identified IT services, that is the IT deliverables made in order to create business value for the organisation. Such services could be classified according to the widely adopted ITIL standard (Information Technology Infrastructure Library). This standard defines the development, deployment, operation and improvements of services as a lifecycle process (see figure 7). The process is defined through the following phases:

Service strategy: This phase is about choosing a strategy which will define the services needed in order to create maximum value with maximum efficiency for the organisation. The service strategy phase is the motor of the service lifecycle and drives all the other phases. The objective is to develop the capacity to gain competitive advantage for the company through the delivered IT services.

Service design: This deals with design and budgeting of the different services. The input is requirements from the customers such as users of the services, and the deliverables are designed services ready to be built and deployed in the organisation.

Service transition: Following on from the design, this phase is about building, testing and deploying the services in the organisation. The service transition is effective and efficient if the transition delivers what the business requested within the limitations of the budgeted resources as defined in the design phase.

Service operation: Concerns the day-to-day operation and servicing of the deployed services. The operation should be according to agreed Service Level Agreements (SLAs) so that the business side knows what to expect at all times. As a natural part of this, measurements of how to monitor performance and stability of the services provided is a natural part of this phase.

Continual service improvement: The delivery of services is a continuous process, and services should be subject to assessments both in respect of how they perform, what they cost and what level of value they deliver to the organisation. Such assessments may lead to improvements or decommissioning of services in the event that the value proves not to justify the cost).

Fig. 7. The ITIL service lifecycle

A key to successful IT performance is the optimal investment, use and allocation of resources including people, applications, technology, facilities and data within these five phases of service provision in the IT organisation.

Such resources could be provided internally by the company or externally by partners and service providers. The decision as to where and how to source the resources is not straightforward, and must as a minimum take due consideration to the following:
–   internal expertise
–   internal capacity
–   security and confidentiality
–   flexibility
–   preservation of business knowledge
–   required service levels
–   outsourcing availability
–   costs.

The primary driver towards outsourcing certain parts of the infrastructure and/or parts of the business processes has traditionally been to save costs. Such cost savings have been identified through so called 'core vs. context analysis'. Core activities are those which a firm must concentrate their own talent, management, and internal resources on, because they are central to the enterprise strategy. Context activities, by contrast, are those that might be reasonably provided to the firm in partnership with other firms for whom those activities are 'core'.

For companies in the financial industry, it has therefore been intuitively appealing to classify IT infrastructure as 'context' and hence an ideal area for outsourcing. The main counter arguments have been in respect of security and stability. As IT becomes the backbone for support of all business processes and reporting, is it prudent to leave the responsibility for stability of this backbone to others? Although these concerns may turn out to be showstoppers, they can in many cases be accommodated through Service Level Agreements (SLAs) and agreements on the right setup – and may therefore be a good solution for some companies.

Outsourcing of business processes is a different matter because these processes represent the manifestation of business knowledge. The organisation therefore needs to have a clear view of the following questions:
– how important is it to have internal knowledge about these processes in the organisation?
– how flexible is the service provider in tailoring their services to my current and future needs?
– in the eyes of the service provider am I a big client or a small fish?
– how easy is it to claw these processes back in house if the service provider fails to provide services according to the SLA?
– how time-critical is the data delivered by the service provider to my overall business flow?
– how important is it that the service provider's calculation methods corresponds with my organisation's own calculation methods?
– how important is it that the service provider uses the same data-sets for these calculations such as prices, yields and FX rates as we do ourselves in our own calculations?

In the asset management industry, there was an increasing trend in the early millennium towards outsourcing of back office operations and associated add-on services such as performance and risk calculations. Many such projects have however failed due to lack of clear answers to the questions set out above, and some asset managers have started clawing the outsourced business processes back in house. The main argument, as explained by these asset managers, was that these operations and services turned out to be 'core' rather than 'context'. This is because quality, flexibility and timeliness of these services are fundamental for the remaining part of the core business. Internal control therefore turned out to be necessary in order to offer satisfactory internal service levels for the business. Such service levels can only be achieved through a stable enterprise data management (EDM)[12] foundation and thereby increased data quality, operational resilience and reduced operational risks.

The bottom line when it comes to sourcing of resources is therefore, when analysing costs and alternatives, to start the investigations by identifying core business areas and then to be very clear about the price one may have to pay in case something mistakenly is classified as 'context'.

Performance measurement

This is the final step in the ITGI model for IT governance. Performance measurement is of course linked tightly to the definition of value delivered by the IT organisation. Some of these target values are measurable, some are more subjective in their nature such as ease of use, look and feel and so forth.

The ITGI recommends the use of balanced scorecard techniques to measure performance of the IT organisation against the identified goals. Balanced scorecards translate strategy into documented action taken in order to achieve the goals, including assessments of these actions for performance measurement. The method provides a tool to measure those relationships and knowledge-based assets necessary to compete in the information age: customer focus, process efficiency and the ability to learn and grow.

The balanced scorecards should be designed to answer questions about the enterprise's way of doing business from four different perspectives:

- **Financial perspective**
  To satisfy stakeholders what financial objectives must be accomplished?
- **Customer perspective**
  To achieve financial objectives, what customer needs must be served?
- **Internal process perspective**
  To satisfy customers and stakeholders, which internal business processes must be excellent?
- **Learning perspective**
  To achieve goals, how must the organisation learn and innovate?

The balanced scorecard approach ensures that performance of the IT function is measured against business values and thus supports the value creation process as depicted in figure 5.

---

12   Schröter, Marc. 2009. 'White paper on Extended Enterprise Data Management'.

IT as a driver for improved GRC in financial institutions

The previous sections have highlighted the fact that governance issues exist for companies in all segments, but also that there are challenges which are particularly pertinent to the financial markets. These challenges have become particularly clear in light of the current financial turmoil. We have furthermore discussed general guidelines to enforce a good corporate governance culture and also touched upon general guidelines as to how IT can contribute to supporting sound governance practice of a company.

This section will focus on some particular problem areas faced by the financial industry and propose suggestions as to how the IT infrastructure can be targeted towards a solution to these problems. Following on from the arguments presented in the previous sections, the main focus areas will be risk and compliance issues.

The fact that IT can make a big difference for financial firms in their endeavours to improve on governance, risk and compliance issues is beyond doubt. A survey conducted by ITGI and PWC in 2008,[13] on a population of 'non-IT related' C-level executives, revealed that half of the respondents indicate that IT is 'very important' to the enterprise. Around 75% affirmed that IT creates business value and the same number considers IT governance an integral part of corporate governance. However, three quarters of respondents felt there were barriers that prevented full returns on IT investment, and these barriers were mainly to do with immature IT cultures and failure of application delivery. This conclusion is however mainly based on subjective assessments, as more than half of the companies surveyed had no procedures in place to measure value generated by IT.

For those who master the challenges, there are however big potential gains on the horizon. In November 2008, Fitch Ratings upgraded the London based asset management firm Schroders Investment Management from M2 to M2+. One major reason for this upgrade was an assessment of Schroders' IT infrastructure. Schroders had recently undergone a project aimed at consolidating their enterprise data into a common 'portfolio book of records', and this new platform gave rise to significantly improved operational as well as financial risk management within the organisation. Fitch Ratings noted:[14]

"The rating also factors in the extent of Schroders' research framework and the solid risk management framework. The addition of the '+' modifier emphasises strength in the company's investment infrastructure and the technological platform with notable progress being made in data management and integration ..."

---

13   'An Executive View on IT Governance'. 2009. ITGI & PWC.
14   Fitch Ratings on Schroders' upgrade.
     http://www.easybourse.com/bourse-actualite/marches/press-release-fitch-upgrades-schroders-invest
     ment-567397.

Enterprise data management

"If you want to build a cathedral, you have to start with a solid foundation"

Schroders' portfolio book of records solution is based upon the simple principle that quality and timeliness of core data are the very prerequisites for any operation in a knowledge-based company. This view seems to be shared by the industry. In a survey conducted by SimCorp StrategyLab,[15] 47% of the respondents rated the following question as 7 or above on a scale from 1 to 10 (1: 'Not at all', 10: 'To a very large extent'):

"The ability to store, process and report high quality, real-time data on risk exposure calculation will influence financial institutions' market value in the future."

Only 13% indicated the issue was irrelevant, rating the statement as 4 or below.

The result is not surprising. The financial services marketplace lives and breathes on information, and according to Russell Ackoff, acclaimed systems theorist and professor of organisational change, the definition of 'information' is *'data that are processed to be useful,'* (providing answers to 'who', 'what', 'where', and 'when' questions). So the conclusion must be that even the best and most sophisticated data processing and decision support tools will fail to provide useful information if the data fed into these tools fail to have the appropriate completeness and quality.

It has already been argued that the likelihood of successful risk management and compliance controls in a company is rather limited without sound corporate governance. This likelihood becomes literally non-existing if the IT infrastructure is unable to provide timely and good quality data across the different operations of the organisation. This is why sound IT governance should start with addressing consistent EDM in the organisation, in order to ensure that the tools and applications available actually provide the employees with the information they need in order to make the right business decisions and thus enable the company to achieve the business goals.

Marc Schröter elaborates on the importance of a sound EDM concept in his 'White paper on Enterprise Data Management'. The white paper discusses different approaches to EDM, from infrastructures where EDM is literally non-existent due to a patchwork of disparate local systems, all with local databases, scattered around in the different departments of an organisation, and through to different degrees of data consolidation in data warehouses, to the ultimate extended enterprise data management (EEDM) concept where all data are consolidated in one common database which then serves as foundation for tools and applications for all departments in a financial institution.

Although the situation seems to have improved over the past few years, the majority of financial institutions still operate rather immature EDM platforms. Other industries, like retail consumer chains, travel agencies and car manufacturers seem to

---

15   'Global Investment Management Risk Survey 2009'. March 2009. SimCorp StrategyLab.

have much more control on data, processes and logistics which are the main drivers for STP. Anyone can today book a flight on the Internet, look up the expected departure time on-line, check in over the Internet or at a terminal in the airport and walk through security and onto the plane without any manual intervention apart from carrying your own hand luggage.

There are of course several reasons why it's not fair to compare the financial sector with these industries. Size, complexity, confidentiality and volatility are some of them. Nevertheless, one likely reason not to be underestimated is the traditional lack of focus on fundamental cross-enterprise issues within financial institutions' corporate governance policies.

Taking a look at international investment management firms, they're typically organised in front, middle and back office operations. The front office may be further split into analytics, strategy, portfolio management and trading. The middle office could be split into risk, performance and compliance groups and the back office into reconciliation, settlement and accounting groups. The combinations could be many, but the point is that the organisations, due to their sheer size, trading volumes and physical distance between offices, have been split into silos. All silos have been given goals to chase 'alpha performance,' that is a level of performance that exceeds a given market benchmark, in their respective areas. So traders are expected to outperform other traders,[16] analysts are expected to pick better performing shares than the competitors, portfolio managers are expected to outperform their benchmarks by setting the right active bets, the risk and performance teams are expected to be market leaders in sophisticated risk and performance reporting, and so forth. Without a strong corporate governance policy to enforce holistic thinking, the impetus will naturally be towards a best-of-breed culture where the different silos are allowed the mandate to pick their preferred tools out of local needs to facilitate local performance. The challenge of the IT department is then to glue all these different tools and applications together in order to create an operational enterprise-wide infrastructure enabling information to be passed from silo to silo. When mapped out on a diagram, the infrastructures of these firms have looked like an overview of electrical circuits in a nuclear power plant, and it is not atypical to see diagrams revealing more than one hundred different software applications put in place to support the operations of a single investment management firm.

Moving on to a structured EDM platform means taking a more fundamental approach to generating alpha. It means alpha in the different silos is driven by a holistic approach, with the view that most silos, at the end of the day, will create better alphas for themselves if they operate on the basis of timely, complete and correct data compared to what they would achieve with flawed data but functionally rich best-of-breed tools. If there are some silos where this is not the case, that is still a price worth paying

---

16   Traders are usually benchmarked against VWAP – Volume-Weighted Average Price on their individual
      trades – which means that everybody is expected to perform better than average.

as the accumulated effect of EDM will, when you add it up across the silos, still give a much stronger and more stable performance for the whole corporation. This move, however, needs to be taken as part of a deliberate corporate-wide strategy and hence requires strong corporate governance in the firm, including solid IT governance.

The situation is of course never black or white. The point here is to stress that corporate governance of financial institutions first and foremost needs to ensure that the information upon which investment and funding decisions are made is as accurate as possible. It is crucial to be able to answer those 'who, what, where, and when' questions in time. *Who* in the organisation is exposed at *what* risks *where* in the market and *when* do we need to take action? For this, timeliness is key. It doesn't help much if the answer to the when question – as information finally is available – is yesterday. Without accurate and timely information, it therefore makes no real sense to discuss risk management or compliance at all.

Risk management

"I cannot imagine any condition which could cause this ship to founder. I cannot conceive of any vital disaster happening to this vessel."—Captain of the Titanic, 1912

If there is something history has taught us, it is to expect the unexpected. If somebody had told an equity analyst in 2006 that, within two years, Citibank would lose 80% of its market value, it would probably have caused some sarcastic remarks. Nevertheless, that's what happened and the whole marketplace was caught off guard.

It is of course easy to be wise in hindsight, and nobody can be expected to be in possession of that famous crystal ball. With proper, corporate-wide, risk management processes in place, it is however possible for a financial institution to become more resilient to adverse market movements, even extreme ones such as those that we have experienced recently.

In dealing with risk management, a company needs to be very clear as to its true business drivers. What is really driving the business? Which main factors, internal as well as external will drive business results up or down? These factors will of course depend on which business segment a company belongs to and which business model the company follows. Within the financial sector, pension funds will have other business drivers than hedge funds and asset managers will have different business drivers than fund administrators for example.

With the business drivers established, the next step is to identify the key risk indicators (KRIs). These are measures which can be used to monitor (lagging KRIs) and even foresee (leading KRIs) events that may impact the business drivers. The KRIs should be distributed to risk owners throughout the relevant departments in the organisation, and should have thresholds attached in order for the risk owner to have a clear view on acceptable tolerances. The risk owner needs to fully understand the nature of the risk in question, but not necessarily the construction of the KRI itself.

What is important is that the owner reacts on the threshold values. Finally, there should be a corporate model to aggregate the KRIs up to relevant sub-levels and all the way up to corporate level.

Good KRIs are often hard to find in practice, and must in some cases be created. This is particularly true for leading KRIs. The market participants are however accustomed to monitoring signals from the market all the time and use models for how key figures like unemployment rates, consumer spending, credit spreads, yield spreads and other factors may influence the economy in a way which could end up influencing their positions. Anyway, a good KRI will fulfil the following criteria:

– availability
– effectiveness/usefulness
– comparability
– ease of quantification

When aggregating KRIs to group levels, the KRIs in the group should work well together in order to be useful indicators of risk level. The following KRI characteristics should be considered before grouping:

– leading vs. lagging
– frequency of collection
– hard metric vs. more subjective, structured evaluation.

In order for this to work on a corporate level, corporate governance should drive a clear and structured risk management policy, including how risk should be dealt with by all levels in the organisation. Risk management must be a focus area for the board of directors and executive management should be actively involved in promoting and incentivising risk monitoring and risk mitigation strategies at all levels in the organisation. It is important to stress that risk management should not be viewed only in a negative vein as precautious initiatives to prevent potential losses. Through a thorough understanding of the business drivers and corresponding risk factors, an organisation will also understand how to better exploit opportunities and hence make sound business decisions to increase profitability.

Even though the recent SimCorp StrategyLab survey on risk management reveals that there has been a slight tendency to move the risk reporting line downwards in the organisation,[17] the general awareness of the necessity to improve overall risk management in the industry is confirmed by the graph in figure 8.

There also seems to be a general appetite for making investments to improve risk management capabilities as a result of the financial crisis. More than 80% of the respondents think it is not unlikely that investments will be made in building compe-

---

17   The survey shows that compared to risk reporting 'in the past', reporting to board of directors has
     decreased by 5% and to CEO level by 1%, whereas reporting to executive management has increased by
     2% and to senior management by 5%.

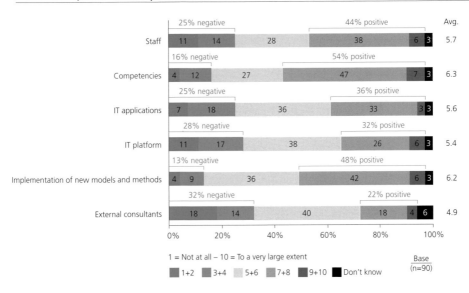

Fig. 8. Areas planned for increased investments in risk management. Source: 'Report on Global Investment Management Risk Survey 2009', SimCorp StrategyLab, March 2009.

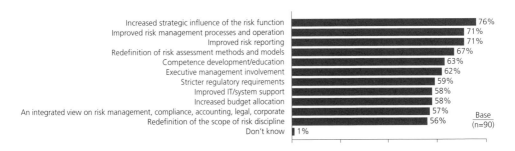

Fig. 9. Keys to improve risk management in the future. Source: 'Report on Global Investment Management Risk Survey 2009', SimCorp StrategyLab, March 2009.

tencies and implementing new models and methods. Furthermore, around 70% did not rule out possible investments into IT platforms and IT applications.

The risk profile of a financial institution can typically be split into the following categories:[18]

– strategic risk
– operational risk
– financial risk
– knowledge management risk

---

18   SimCorp StrategyLab: Report on Global Investment Management Risk Survey 2009, March 2009.

 – compliance risk
 – reputational risk.

In the following, we will cover operational risk and financial risk. Compliance risks will be covered together with other compliance issues in a later section.

Operational risk

Operational risk concerns the likelihood and impact of events occurring which will affect the daily running of the business. This covers a wide range of different events such as:
 – IT disruptions or failure of external service/product providers
 – fraud, both external and internal
 – theft of confidential data and hacking
 – external robbery and theft
 – erroneous models (incl. pricing of holdings)
 – faulty data
 – insufficient transparency in data
 – insufficient dissemination and distribution of information
 – rogue trading and self-dealing
 – errors in legal documents/compliance processes
 – loss of key personnel
 – mis-selling/incorrect advice or failure to assess product suitability.

Many of the typical operational risks sort under the responsibilities of the IT department. It has been mentioned that modern governance of a complex corporation is literally impossible without a sound IT infrastructure aimed at delivering business value for the organisation. This then means that the organisation becomes extremely dependent on stability of the IT services, and that severe disruptions tear away the rug from underneath the whole company. As discussed, this argument could become a showstopper when it comes to a decision as to whether or not to outsource parts of the IT infrastructure. Is it prudent to rely on an external provider for this? Depending on the circumstances, the answer could be yes, but a decision to outsource certainly means that this risk will be harder to control and/or mitigate for the organisation.

The flexibility and robustness of the IT infrastructure will typically come to the test when big changes occur to the organisation, such as in the case of mergers, divestments or internal reshuffling of organisation or business models. Again, such situations have suddenly become very real with all the restructuring taking place as a result of the financial crisis.

IT will also have a significant influence on the organisation's ability to prevent rogue trading and breaches of internal mandates. Everybody remembers Nick Leeson bringing down Barings Bank through reckless derivatives trading on the Asian exchanges in 1995. More recently, French Bank Société Générale suffered a loss of more than $7 billion due to rogue trading through a number of transactions by a single employee. It is fair to accept that no control measure is 100% fail-proof, but one must nevertheless expect from a sensible IT infrastructure and sound reporting procedures that consecutive breaches by single employees should be detected.

One big issue, which became very relevant when the skeletons started rattling out of the cupboards in 2008, was a company's ability to quickly get an overview of total exposure to the different entities in the market. This exercise is not straightforward for a big international investment firm. Exposures are normally hidden through different layers of derivative contracts and other structures as well as scattered around in different funds and portfolios are often recorded in different databases in different geographical regions.

The process leading to the collapse of Lehman Brothers in September 2008 moved very quickly. In a matter of days, stakeholders and creditors had to figure out how to protect their exposures. In a move to help, the International Swaps and Derivatives Association (ISDA) actually opened a special trading session on Sunday 14 September to allow market participants to offset their positions in various derivatives, but it didn't provide much help for those firms which didn't know what their overall positions really were at that point. On Monday 15, Lehman Brothers filed for bankruptcy. Even after that, it took many stakeholders weeks to sort out their overall claims/losses against Lehman Brothers. The problem is of course of a market risk nature, but it also becomes an operational risk due to the operational challenges in collecting and disseminating data from the various contributors to the overall exposure. This is the sort of operational risk which makes it blatantly clear that well-wrought data management is instrumental in the financial sector. It highlights the fact that when systemic risks materialise, the 'silo' approach is not good enough. The different departmental exposures in a firm will have a dynamic effect throughout the organisation which means the sum of local risks do not necessarily add up to the overall corporate risk. An organisation therefore needs to be able to swiftly get the consolidated overview so that action can be taken on corporate level. In this respect, one may argue that the 'best-of-breed' IT infrastructure resulting from the 'silo approach' in the financial industry as discussed earlier is an example of a 'good weather' platform which may produce good alpha in normal and stable markets, but which represents significant operational risk for the organisation. This risk will surface in times of systemic crises when departmental thinking is insufficient and a holistic view is paramount.

Some examples of KRIs pertaining to stability, flexibility and security of the IT infrastructure could be:

Stability
- statistics for central processing unit (CPU) usage
- statistics for network traffic
- statistics for database traffic and available space
- statistics for how fast the databases grows, taking archiving options into consideration
- statistics for downtime of applications in the infrastructure
- statistics for response times of applications
- statistics for recovery time in case of crashes.

Flexibility
- statistics on how long it takes to set up a new portfolio or a legal entity
- statistics on how long it takes to add new applications to the infrastructure
- statistics on how long it takes to report total counterparty exposures
- statistics on how much redundant data, that is data which already exist elsewhere, a new application creates in the infrastructure
- statistics on how long it takes to react on key external events.

Security
- statistics on external hacking attempts, both successful and failed
- statistics on internal breach of compliance rules and authorisation levels.

Financial risk

For companies making a living out of investment and funding, it is quite obvious that financial risk represents a major part of the overall risk profile. Financial risk can be split into further categories like market risk, credit and counterparty risk and liquidity risk. In the following, we will take a look at these categories and propose some IT values which can be used to drive the IT infrastructure towards mitigation of these risks.

Market risk

Market risk can be defined as a risk which is common to an entire asset/liability segment. According to traditional capital asset pricing model (CAPM) portfolio theory, market risk is associated with systematic risk in that this risk cannot be diversified out just by picking different stocks in the same segment. Portfolio diversification then needs to take place by picking stocks from different segments.

The Basel II regulation recommends Value at Risk (VaR) methods to measure market risk. This is a statistical method which, based on correlation techniques, gives a likelihood for a maximum loss for a portfolio over a certain time period.[19] VaR can be calculated in three ways:

Historical VaR
- Calculation is based on historical or logged values for all holdings in the portfolio

Parametric VaR
- Calculation is based on current volatilities and correlations according to assumed distribution model between holdings

Monte Carlo VaR
- A number of simulations of historical prices are generated such that these prices satisfy current volatilities and assumed correlations in the portfolio. The resulting simulated price scenario is then used to calculate VaR (as for historical VaR).

It is obvious that such VaR calculations require a strong data foundation with a comprehensive analytical tool on top. Those entities that calculate historical VaR need to maintain full price history, on a daily basis, for all assets in all portfolios. This gives rise to certain challenges when it comes to data quality, its cleansing, consistency and timeliness, but also to verification of the prices in the database. If the model is built and the data is in place, it requires relatively little effort to press the button and start a VaR calculation. To verify the results, however, becomes a needle in a haystack exercise and is literally impossible unless the VaR tool contains good tracing features which also helps documenting where the prices come from such as from upload from an external data vendor, manually keyed, theoretically calculated and so forth.

For those who rely on parametric VaR or Monte Carlo Var, the requirement for historic prices is less, but many of the challenges in tracing are still the same. These methods are reliant on upload of volatilities and returns from external data vendor, and this obviously brings forward some requirements to connectivity and data import.

Although VaR techniques are commonly accepted as useful indicators for market risk, they are also associated with obvious shortcomings. As with any statistical method, they can only say something about risk levels when markets are relatively stable. It may well be the model says there is only a 5% likelihood of losing more than $1million, but what if you end up in that 5% interval? What amount of money do you risk losing then? This is why statistical methods like VaR should always be followed up by stress testing scenarios, and always be used with care in times of extreme price movements. Such precautions are of course even more important if such movements happen in segments to which a portfolio/company is heavily exposed.

An obvious example to pick is the financial institutions exposure to the real estate segment. It is not surprising that price movements in the housing market is a major KRI for economic activity in general. After all, housing expenses are by far the big-

---

19   For instance, if the three days, 5% VaR of a portfolio is $1m, it means that there is a 95% likelihood that the portfolio will not lose more than $1m over the next three days.

gest expense for an average family in the industrialised world, so movements in these prices will significantly decide buying power and hence the community's capacity to fuel the economy. Nor is it surprising that real estate, for the same reasons, ends up as the main collateral for any sort of funding, including all sorts of investment products originating from funding arrangements.

What is surprising though is how the sector failed to take this KRI seriously. According to traditional asset evaluation theory, the price of an asset should correspond to the net present value of future cash flows generated by the asset. In the housing market this cash flow would correspond to realistic rental income for properties. So there ought to be a balance between the price level for purchase of properties and the price level for rental of properties. This relationship is expressed as the price-to-rent ratio, and depicted in figure 10. The ratio shows that property prices in the early parts of the millennium increased sharply compared to the rental level, hence indicating the build-up of a bubble in the housing market. This curve peaks in the early parts of 2006, where house prices, over six years, have risen almost 80% compared to a stable, balanced level to rental prices.

Now could such over-valuation of properties be justified by the argument that the society was moving towards a new economy where these new price levels would be sustainable? It is hard to see how this argument could be justified. The only explanation would be if the expectations were that renters would be willing/able to pay significantly higher rents in the future, and such expectations must depend on corresponding expectations of household economies. Figure 11 shows the development

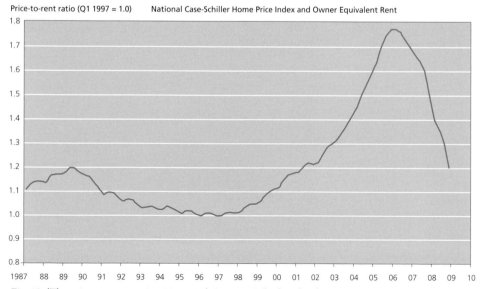

Price-to-rent ratio (Q1 1997 = 1.0)   National Case-Schiller Home Price Index and Owner Equivalent Rent

Fig. 10. 'The price-to-rent ratio 1987-2009'. Source: CalculatedRisk, 26 May 2009.
http://www.calculatedriskblog.com/2009/02/house-prices-real-prices-price-to-rent.html.

of house prices compared to household income. Again, we see a similar imbalance. Household income did not keep up with the increasing property prices, indicating that there would be no room in the household income to pay rents at a level that could justify the pricing boom.

These strong indications from basic property market KRIs make it even more surprising that banks then failed to control the increasing overexposure to this overheated segment. The banks, already exposed to the sector by some 40-60% of their balance sheets, geared up this exposure by investing in securitised loan arrangements like CDOs, which had a significantly lower capital charge compared to typical traditional loans. In early 2008, the US banking sector held $1.3 trillion worth of securitised loans. To top it up the banks geared up their exposure by offloading these instruments to SPVs, but at the same time retaining the financial risk through provision of liquidity and credit enhancements towards these SPVs. Such enhancements are treated as even more capital light than the offloaded CDOs, effectively allowing five times higher leverage ratios compared to what would have been allowed if these instruments were kept on the balance sheet. The effect is the same as writing a put option to the market – "if the SPVs fail to make their commitments, we still guarantee your investment".

Another structure that went out of control was credit default swaps (CDS). This is basically an insurance policy against credit default events. The buyer of protection pays a regular fee to the seller, an insurer, until maturity of the swap. The seller will, on the other hand, pay an agreed amount in case of an agreed credit event during the lifetime of the swap. The CDS market in the US grew from $631 billion in the first

Price to household income (Q1 1987 = 1.0)

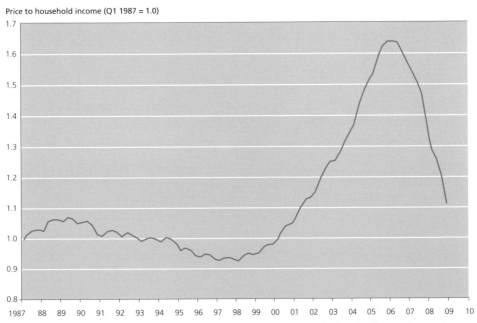

Fig. 11. The price-to-household income ratio 1987-2009. Source: CalculatedRisk, 26 May 2009. http://www.calculatedriskblog.com/2009/02/house-prices-real-prices-price-to-rent.html.

half of 2001 to $54.6 trillion in the first half of 2008. These CDSs were written against credit events in the housing market as well as in the corporate bond market, and most of them ended up on the books of banks and hedge funds. The rest is history. The credit events occurred, and gave rise to huge losses for the insurers.

These examples lead to one other interesting observation about the financial marketplace – its cyclical nature. As mentioned before, there have already been two 'crises' in this century, although for different reasons, and going further back in time there could be good arguments that the financial markets typically will fluctuate between peaks and troughs in 6–10 yearly cycles. One good explanation for this could be the immense innovation and attention level in the industry. Somebody will always find new cunning ways of making money and as soon as the news spreads, everybody wants to share a piece of the success and jumps on the bandwagon. This will have a self enforcing effect which will push up prices all the way until the fundamentals, for instance house prices and income levels, cannot sustain the trend anymore. This will then also mean that downturns will always be abrupt. A bubble does not peter out – it explodes. It also means that 'good weather' infrastructures are not sustainable in the long run. When the bubble bursts, the winners are those who already – before the explosion – had a good understanding of its size and who had an exact and consolidated view on their exposures.

The crux of the matter is, again, transparency and overview across the different business lines. Some crucial questions the banks must ask themselves are;
– did we, at any point, have a consolidated picture of our total true gearing levels?
– did we, at any point, establish proper KRIs associated with this exposure, local as well as consolidated KRIs?
– did we, at any point, have risk mitigation strategies in place in case the KRIs exceeded their threshold values?
– did we, at any point, have a consolidated picture of our true total dollar exposure to the housing market through loans, CDOs CDSs, SPVs and other structures?
– did we, at any point, have a true understanding of the mechanisms which would convert our exposures to true dollar liabilities on our side?
– did we, at any point, have a measure as to what these liabilities could be?

Those who can answer 'yes' to all of these questions are probably the ones who have suffered the least during this crisis. They must have been able to see what they really invested in, and been able to consolidate this across the enterprise and take appropriate measures at the right levels in the organisation when the alarm bells started to ring. For this, they probably have the following infrastructure in place:
– a sound corporate governance policy focused on business drivers and delegation of risk management responsibilities;
– a sound data structure providing full look-through on structured products and

layered investments such as options, futures, swaps, repos, CDOs, fund units and other assets;
–   strong analytical tools able to split investment products into their different basic components and value these components individually;
–   strong reporting tools able to process the data and provide the right information at the right time to the right levels in the organisation;
–   a well-designed EDM platform for timeliness and consistency of data across the entire enterprise.

So in times of extreme imbalances in the market and/or extreme movements in prices, the message is: 'back to the basics'. No sophisticated mathematical risk calculation will be able to save you. Any risk management policy should therefore ensure sound and basic transparency on the basic bread and butter risk exposures in addition to the sophisticated mathematical methods. A common prerequisite for both is though a sound EDM structure across the enterprise.

Credit and counterparty risk
Credit risk and counterparty risk are terms sometimes used interchangeably in the financial markets. For this article, this type of risk will be defined to cover the following eventualities:
–   risk related to an asset's credit rating which may deflate the market value of this asset;
–   risk due to uncertainty in a counterparty's capability of meeting its obligations in respect of an agreed trade.

This type of risk concerns solvency and companies' ability to meet their short-term commitments. Such ability is of course crucial in the marketplace's evaluation of a company. Credit and counterparty risk are therefore relevant when evaluating risks of falling market value of holdings, as well as risks of not receiving outstanding payments from counterparties in respect of agreed transactions.

Listed companies and their funding instruments such as shares and bonds are credit-rated by the international rating bureaux such as Moody's, Standard & Poor's and Fitch Ratings. These credit ratings will impact on the risk adjusted value attached to shares, bonds and loans issued to these companies and hence impact on the allowed leverage for banks and other institutions, subjected to capital solvency requirements, holding such assets on their balance sheet. Furthermore, credit ratings are highly relevant for asset and fund managers in that many investment products have mandates set out for the portfolio manager in respect of the credit ratings of the assets in the portfolio, for example 'must contain at least 25% AAA rated US shares'. Furthermore, sound portfolio management practice should seek to minimise the risk for a given target return, and this means diversification of holdings across companies and asset classes. Keeping full control on the portfolio's exposure to credit risk is instrumental in this respect.

On the trading desk, good governance practice should reduce overexposure to trading counterparties. This could be enforced by defining maximum credit limits to the different counterparties for the different traders and for the trading desk as a whole. Whenever a counterparty is to be picked for a trade, the trading system should then automatically advise as to which brokers are available out of considerations like maximum allowed trading amount for the dealer and remaining credit amount with the broker.

Market risk and counterparty risk are highly interlinked as companies' creditworthiness, to a great extent, will be driven by how they're affected by different kinds of market movements and events. Particularly when it comes to systemic risk, it is obvious to see the strong relationship between the two types of risk. The systemic risk that led to the downturn in the financial markets also created a general scepticism and lack of trust to counterparties and whether these would live up to commitments. This then led to a reluctance to invest and lend money, which again led to the so called 'credit crunch'; companies were exposed to liquidity risk.

The main point to make in this section is again to highlight the importance of knowing your true exposure at any point in time, including the duration of this exposure and a measure of risk associated with the exposure. How quickly can you react if you need to unwind your investments with a company and claim your outstanding entitlements? What sort of engagements do you have with this company and where are you in the priority chain if the company goes bust?

Liquidity risk
"Cash is King"

In general, liquidity risk is about the risk of not having enough cash or short-term funds to meet daily obligations. In finance this risk extends to the risk of not being able to convert financial assets into liquidity due to lack of buyers in the market. If, for example, a bank wants to reduce its exposure in a certain share, and there are not enough buyers to buy this amount of shares, the bank is exposed to liquidity risk. They will need to sell the position in pieces, and risk that their activity of selling a large position in the share in itself will drive the price down. So liquidity risk in this sense will have an adverse effect on the average price achieved for the position and hence result in less liquidity as a result of the sale, compared to if there were enough willing buyers for the original price.

As lack of liquidity is so noticeable by the market, liquidity risk is a factor that often both signals and triggers other types of risks. In the run-up to the collapse in

2008, two highly leveraged hedge funds managed by Bear Sterns essentially collapsed. These hedge funds were mainly invested in sub-prime asset backed securities. As the value of the collateral started to decline with the falling house prices, see figure 10, lenders to the fund started to demand more collateral. One of the lenders, Merrill Lynch, in fact seized $800m of the fund's assets and tried to auction them off. They could only sell $100m and there was no appetite in the market for the rest. When the market realised this, it in fact set off the downward spiral for all related instruments such as CDOs, CDS, and others and triggered the beginning of a systemic crisis.

The example highlights the importance of available liquidity in the market. Lack of liquidity is poisonous for the financial market, and when it spreads to become a general issue across asset classes, the sector will typically find itself in a self-enforcing spiral as depicted in figure 12. Companies with liquidity constraints are forced to sell off assets. This will drive the price of assets down, which again will cause companies to be in breach of solvency rules. This will then make funding more difficult and even force further sales of assets.

Fig. 12. The liquidity spiral in financial markets

One other reason for paying special attention to liquidity risk is the limited possibilities for hedging. It is, for example, possible to buy some level of insurance against drops in assets values, that is market risk, or against credit risk events but this will not necessarily remove the liquidity risk. Liquidity risk must therefore be controlled by sound asset/liability directives and monitoring as well as succinct cash forecasting based on accurate and consolidated data.

Compliance

"You do not get what you expect – you get what you inspect!"

The term 'compliance' concerns a company's adherence to controls, limits and thresholds set out by internal and external stakeholders. The financial industry is heavily regulated, and discussions on compliance often revolve around how the different entities best can comply with international and national regulations. Failure to comply with these requirements will expose a firm to reputational risk, fines, or even loss of license to operate its business.

To be on the right side of laws and regulations is of course instrumental, but one must not forget that industry regulations only set out the minimum requirements a firm has to abide by. Furthermore, such laws and regulations are generic to an entire industry sector, and can never cover every aspect of the true risks in this sector. Companies with sound corporate governance policies therefore do not confine themselves to compliance with external directives, they supplement with internal controls to supervise and monitor internal rules set out to control identified risks (KRIs) to their own business drivers as discussed earlier. It is therefore worthwhile to distinguish between external and internal compliance. External compliance is about abiding by regulations and expectations set out by external stakeholders, whereas internal compliance concerns living up to the company's own internal policies.

One thing that distinguishes internal and external compliance is the reporting level in the organisation. External compliance normally requires for a legal entity (a company, a fund, or a trust) to report the consolidated numbers for the entire entity as such to the relevant external stakeholders. Internal compliance is part of the governance policy, and will involve local compliance at all different levels in the organisation. As an example, we've already mentioned Basel II for the banking industry and Solvency II for the insurance industry. These regulatory standards set out to control the banking and insurance industry respectively in respect of:
- risk management and minimum capital requirements or solvency
- effective supervision by regulators or external compliance controls
- disclosure requirements or external compliance reporting.

As mentioned, the capital requirement directives set out rules for how capital should be set aside' to provide for risks associated with assets on the balance sheet. This capital provision reduces the allowed gearing level and is therefore a measure of the price a bank or an insurance company has to pay for keeping these assets on their balance sheet. Sound governance policy will introduce internal rules that allocate capital to the relevant value generators in the company. So risk managers, portfolio managers, analysts, traders and others will all have their share of this 'cost'. If the cost is built into their performance measurements and hence bonus schemes, there will be an incentive to manage it and keep it at a minimum level.

Furthermore, this cost model should be extended beyond the minimum rules set out by Basel II and Solvency II. Banks, insurance companies and others should have cost models in place that reflect the true risks these companies are exposed to. This involves exposures through the balance sheet as well as outside the balance sheet. In respect of the already mentioned SPVs set up to host CDOs, the banks should ask themselves: how will the issued credit and liquidity enhancements hit us if the SPV goes bust? And even if we haven't issued any credit lines, do we really want to run the reputational risk that comes with letting investors lose out in a major way on a product under our name? If the answer to these questions are that the bank will suffer the loss anyway, then this risk should be priced just as if the instruments were on the balance sheet. The point is to not hide anything under the rug, but to be candid and honest in the identification of the true risk the company is exposed to. The 'tone at the top' as explained with figure 1 needs to send this strong signal all the way through all levels in the organisation. The true corporate risks should be given cost tags and allocated throughout the organisation. If a sensible model can be developed to calculate how this cost tag should be priced at each level in the organisation, then the tag can be used as a KRI. The company can then implement compliance controls on thresholds for each allocated cost.

The aim must be to introduce compliance controls that protect the business values and create stability rather than only satisfying the regulators. The approach discussed here requires:

- a good corporate governance policy aimed at enforcing a stable and transparent business model and not invented for regulatory arbitrage;
- solid analytical tools to provide a full overview and pricing of all assets;
- a comprehensive accounting tool to calculate booked positions and profit and loss (P/L) according to the different accounting rules a company needs to abide by, for example IFRS, tax accounting and local GAAP;
- a sensible evaluation model to calculate risk costs at all levels;
- strong reporting tools able to report risk costs against risk budgets and thresholds for costs at all levels in the organisation;
- a solid and comprehensive EDM foundation.

In respect of the second bullet above, it is important to bear in mind that an analytical tool should be able to price assets theoretically even if there are available market prices for these assets from external price providers. As one initiative to prevent the liquidity spiral as depicted in figure 11, the legislators discuss the possibility of opening up for internal pricing of assets in case liquidity is drained from the market. This could allow companies to keep these assets on their balance sheets rather than being forced to sell them in a market which is out of balance. This discussion has not yet been concluded, but if it materialises it means the legislators will require sound economical models as foundation for any theoretical price which has replaced a market price for evaluation of assets on the balance sheet.

Furthermore, also in support of quality data and strong analytical tools, compliance with Basel II requires VaR calculations for market risk and also potentially sophisticated analytics for reporting of credit risk. According to Basel II, credit risk can be approached in one of the following three ways;

1 standardised approach
2 foundation, internal ratings based (IRB) approach
3 advanced IRB approach.

The standardised approach entails risk weighing of assets based on standard ratings from rating agencies. The two latter approaches (IRBs), means that banks, if they have the required credit rating tools available, are allowed to rate the assets internally and this will normally give lower risk weights compared to the standardised approach. Many banks are therefore inclined to develop such tools and report credit risk according to foundation IRB or advanced IRB approaches. These approaches, however, are based on sophisticated distribution models and statistical methods.

The term 'compliance risk' is normally used for risks pertaining to the consequences of not complying with regulations and mandates set out for the entity, the company, fund, trust and so forth. As mentioned above, this can have severe financial consequences. Fines and other punitive actions imposed by regulators as well as compensation claims from clients can be bad enough, but the worst financial consequences normally come through reputational risk, that is the risk of losing clients and market share due to a dented brand.

With the model discussed in this section, it will also make good sense to define internal compliance risk. This kind of compliance risk will then pertain to all units in the organisation with a risk budget, and the compliance risk will be the financial impact on the organisation if the unit exceeds this risk budget.

Conclusion

This article has discussed issues relating to the financial industry in respect of governance, risk and compliance. To illustrate the issues, references have been made to the current financial crisis in the worldwide economy. It has not been my purpose to analyse this crisis in detail, but rather use the examples as input to some proposals on how IT can contribute to supporting a better corporate governance practice in the financial sector.

This practice must be driven by top management, and starts with the board of directors and executive management defining a clear policy for sound corporate governance. This corporate governance policy needs to also cover an IT policy, which is aimed at delivering value to the business drivers in the organisation. The governance policy should ensure risk measures are introduced at all levels and quantified as risk budgets. Compliance controls should be implemented to report on how the different internal units perform against their risk budgets.

It has been illustrated that the financial industry is heavily reliant on information, and this information needs to be of good quality and delivered at the right time to the right people. For any company, the flow of information goes both ways. The company needs access to information from the markets and it also needs to contribute with timely and precise information to the market about its own business performance, including its risk profile. The basic ingredient for information is data. Financial institutions therefore need to improve their EDM models, which traditionally have suffered due to departmental silo thinking. The second ingredient is analytical tools, which are able to transform the data into the information needed in order to report risks and control compliance on departmental levels as well as on a consolidated corporate level. Finally, a strong reporting structure is needed in order to accumulate information and pass the right level of information to the right levels in the organisation.

The message is therefore to start with the fundamentals. If the right foundation is in place, then sophistication can be increased over time as a controlled process. If the foundation is not of appropriate quality, further sophistication will probably just exacerbate an already fragile information infrastructure. The extra information provided with the increased sophistication will therefore at best be of little or no value, and at worst be directly misleading and introduce further risks for the enterprise. Boards and executive management of global financial institutions are faced with challenges comparable to sorting out a giant jigsaw puzzle. Sound corporate and IT governance will ensure all the pieces are on the table and the overview is provided in order to get the pieces together and get the full picture. In this respect, a solid EDM platform is key in any modern financial institution.

References

Acharya, Viral & Richardson, Matthew (eds.). 2009. 'Restoring Financial Stability'.
New York University, Stern School of Business.

'An Executive View on IT Governance'. 2009. ITGI & PWC.
http://www.pwc.com/Extweb/pwcpublications.nsf/docid/A9EC7D4142D2102A85
2575AA000D91D0.

'Board Briefing on IT Governance', 2. edition. 2003. ITGI. http://www.itgi.org/Tem-
plate_ITGI.cfm?Section=ITGI&CONTENTID=6658&TEMPLATE=/Content-
Management/ContentDisplay.cfm.

Fitch Ratings on Schroders' upgrade.
http://www.easybourse.com/bourse-actualite/marches/press-release-fitch-up-
grades-schroders-investment-567397.

'McKinsey Global Investor Opinion Survey on Corporate Governance'. 2002.
McKinsey.
http://www.mckinsey.com/clientservice/organizationleadership/service/corpgov-
ernance/pdf/globalinvestoropinionsurvey2002.pdf.

Moorman, Theodore Clark. 2005. 'Corporate governance and long-term stock returns'. Dissertation submitted to Texas A&M University, August 2005.
http://txspace.tamu.edu/bitstream/handle/1969.1/2341/etd-tamu-2005A-FINC-
Moorman.pdf;jsessionid=6E1FDFE001F532ED5C21C82445DBFFE8?sequence=1.

O'Donovan, Gabrielle. 2007. 'The Corporate Culture Handbook: How to Plan, Implement, and Measure a Successful Culture Change'. The Liffey Press. (http://txspace.
tamu.edu/bitstream/handle/1969.1/2341/etd-tamu-2005A-FINC-Moorman.pdf;js
essionid=6E1FDFE001F532ED5C21C82445DBFFE8?sequence=1).

'OECD Principles of Corporate Governance'. 2004. OECD. http://www.oecd.org/
dataoecd/32/18/31557724.pdf.

'PHLX KBW Bank Sector Index, Sep 2008–May 2009'.http://finance.yahoo.com.
http://www.analyzeindices.com/ind/banks.htm.

'Report on Global Investment Management Risk Survey 2009'. March 2009. SimCorp
StrategyLab.

'The Risk and Control Framework Model'. 2008. Ernst & Young.

Schröter, Marc. 2009. 'White paper on Extended Enterprise Data Management'.
SimCorp.

In practice

11 Three executive interviews: evidence from the front line of investment management

by Caspar Rose

Introduction

This chapter, based upon interviews with key representatives from three different leading financial institutions, aims to provide a better understanding of the causes behind the current financial crisis as well as its consequential changes to risk management in these specific institutions: Deutsche Bank, Allianz SE and ATP, the largest Danish institutional investor. Three very senior, experienced individuals have been chosen for the interviews as they have all witnessed at first-hand the strategic financial decisions which prior to the crisis were taken in large international financial institutions. Moreover, they all contribute valuable insights as to the causes of the current crisis as well as recommendations about how to avoid another similar crisis.

The first interviewee is Tom Wilson who has been Chief Risk Officer at Allianz SE since 1 January 2008. Wilson was previously Chief Financial Officer and Chief Risk Officer at Swiss Re New Markets, the capital markets and alternative risk transfer division of Swiss Reinsurance Co. Tom Wilson has also been Global Head of Risk Management Practice at McKinsey & Company. Recognised as 'CRO of the Year' in 2007 by Life & Pensions magazine, Wilson holds dual Swiss and American citizenships.

Chief Risk Officer at Deutsche Bank Dr. Hugo Banziger, the participant in the second interview, became a member of the Management Board at Deutsche Bank on 4 May 2006. In 1996, Banziger joined Deutsche Bank in London as Head of Global Markets Credit. He was appointed Chief Credit Officer in 2000 and became Chief Risk Officer for Credit and Operational Risk in 2004. Dr. Banziger began his career in 1983 at the Swiss Federal Banking Commission in Berne. From 1985 to 1996, he worked at Schweizerische Kreditanstalt (SKA) in Zurich and London, first in Retail Banking and subsequently as Relationship Manager in Corporate Finance. Dr. Banziger is a member of the Supervisory Board of EUREX Clearing AG, EUREX Frankfurt AG and a member of the Board of Directors of EUREX Zurich AG.

The third interview is with Lars Rohde, since 1998 CEO of the largest Danish Institutional Investor named ATP (Arbejdsmarkedets Tillægspension). Rohde has a background as an economist and has worked at the Danish National Bank as well as holding leading positions in a Danish pension fund (Lægernes Pensionskasse), from 1988 as CEO. Subsequently, he became deputy CEO in Realkredit Danmark, the largest Danish real estate finance fund. Lars Rohde was a member of the supervisory board of the Danish Stock Exchange during 1993–1996 and is currently a member of the Danish Corporate Governance Committee. Lars Rohde is also an external associate professor at the Copenhagen Business School.

Indentifying the causes of the financial crisis – what went wrong?

This article reflects the views of Tom Wilson, Chief Risk Officer, Allianz SE.

The causes
According to Wilson, the current financial crisis was caused by the following:
- exaggerated leverage
- insufficient regulation of capital markets
- myopic shareholders
- unhealthy public policy.

Prior to the current crisis, the US Government stimulated home ownership and supported people's desire to own their own homes. Financial institutions competed for borrowers and credit to households as well as firms became cheaper to obtain. As a consequence, a housing boom was created and fuelled by aggressive mortgage lending, including substantial sub-prime mortgages. Much of this lending was made carelessly, often without taking into account whether or not the borrowers could afford their monthly payments.

Retrospectively, it appears that lenders and consumers acted as if they had never experienced a crisis before. This is turn led to numerous poorly conceived loans being funded by the lending banks through the wholesale financial market by means of collateralised debt obligations (CDOs).

Moreover, current regulatory controls proved to be insufficient as regulators did not step in in time to prevent the situation developing into the crisis that it became. The lack of a single effective financial authority is apparent. Also, more transparency will be needed in the future to enable outsiders to determine the risk exposure of financial institutions. Deposit insurance laws are in place in most jurisdictions, but contrary to the common view, Wilson does not believe that a deposit insurance system poses any substantial moral hazard.

Wilson also criticises investors as being one of the causes of the crisis. Myopic shareholders increasingly expected higher returns which led to systemic failure because in order to generate these high returns, asset managers were tempted to focus on the short-term profits, while ignoring the long-term risks.

Another reason for the crisis, was the loose monetary policy applied by the US Federal Reserve (FED), which did not pay attention to the developing risk of a crisis in due time.

Herd mentality

The banks played a key role in triggering the financial crisis when setting up off-balance sheet 'warehouses' to hold various financial contracts. The largest banks all wanted to take the lead by offering a huge variety of financial products, including CDOs, which had become tremendously popular in the years preceding the crisis. All the major banks were involved in massive securitisation stimulated by short-term incentives offered to their asset managers. The excessive leverage raised by the banks, in order to fund their portfolios of assets, in combination with the use of off-balance sheet structured investment vehicles (SIVs), contributed negatively to the crisis.

All the large banks wanted to be in the lead, so if a bank designed a new structured product all the other banks followed shortly afterwards with similar products. This herd behaviour was evident even when the banks' top management did not fully understand the products' risks. "When the music started to play people started to dance" says Wilson.

The Basel requirements are not sufficient

The Basel requirements, which are incorporated in the European Union's Capital Requirements Directive (CRD), deal with solvency requirements, but they were inadequate and did not prevent the crisis or even reduce its negative consequences. SIVs, for example, were not covered by the Basel requirements.

Internal models are valuable tools for risk management, but the inability of the Basel requirements to mitigate a systemic crisis has been proved.

Governance and operational risk

Wilson argues that the crisis has not made management of operational risk more important than before the crisis. However, he says that it is now easier to spot. Moreover, during a crisis people may be tempted to commit fraud or manipulate earnings in order to sustain the same level of income.

Competent people are crucial in managing operational risks effectively. People with substantial financial experience who are capable of asking the right questions as well as taking good decisions are especially important. These could be people from the front office who want to change their career, but still want to capitalise on their knowledge of financial markets and securities transactions.

The board of directors bears ultimate responsibility at a firm, but increasing the board's involvement does not avoid a crisis. Members of the board of directors are often too distant from the operations of the firm. Even with a strong chairman combined with an increase in the number of board meetings there is not much the board can do if the firm is impacted by a crisis. Instead one may argue for the importance of a first line of defence, for instance, underwriters, risk managers or other front line personnel.

On the other hand, Wilson suggests that the board needs to increase the entire firm's risk awareness based on professional advice from the firm's risk experts. He believes that the Chief Risk Officer should be member of the board in financial institutions.

The way out of the crisis

Allianz expects a slowdown in the economy during the next two years, or perhaps a shorter period. Businesses will suffer a decline in orders followed by increasing unemployment and lower consumption. This is a paradox, says Wilson. "It was cheap money that got us into the crisis," he says, "and it is cheap money that will get us out, in a process of structured readjustment". In other words, there is a need for a consumer-lead recovery. The process can be reinforced if the dollar depreciates in value as fewer foreign investors want to hold assets denominated in dollars. This will increase US exports as US goods become cheaper which will, in turn, boost the US economy and eventually the world economy.

A radical, but effective solution to mitigate the crisis is to force liquidation of bad asset securities as well as those financial institutions that hold poor assets. Wilson concludes that the government should not care about the interests of existing shareholders as their investments are already lost. Instead there is a need for a controlled failure where the government creates a new ball game by providing equity to sound financial institutions.

Securing bank interconnections is key to stability

Interview with Dr. Hugo Banziger, Chief Risk Officer and Member of the Management Board, Deutsche Bank.

Recovery linked to US housing market

Generally, commentators have blamed the financial crisis on excessive leverage and short-term incentive contracts. Similarly, the low interest rate, low inflation environment, which led to a prolonged period of benign markets, over-reliance on external ratings, governance weakness and poor risk assessment all dulled investors' perceptions of the risks they were taking. However, it is the scale that separates this crisis from its predecessors. "The collapse of the US housing market made this crisis truly global," says Dr. Hugo Banziger. He believes stabilisation of house prices in the US is an essential factor in restoring confidence and driving the recovery of the broader world economy. "If you don't know how much your house is worth there is uncertainty – and certainty and trust are key to the recovery," adds Dr. Banziger. US house prices have fallen by approximately 25% since the crisis began and, with many potential purchasers still unable to secure a mortgage, economic predictions suggest they could tumble a further 10–15%.

Narrow banking is not the answer

Many have compared the current crisis to that of the 1929 crash and indeed there are many similarities and lessons to be learned. Like back then, stabilising banks, easy monetary policy and the use of public money are all necessary for overcoming systemic shocks. The basic structure of our current financial regulation originated from the 1929 crisis and the investigations that followed paved the way for the 1933 Securities Act and 1934 Securities Exchange Act. These put minimum capital and liquidity requirements, good corporate governance, and disclosure and transparency at the heart of necessary improvements. Notably, the US response to the crisis contained measures (later repealed in 1999) to separate commercial lending activities from investment banking – a move that some are now calling for a return of in the wake of the recent crisis. However, Dr. Banziger sees no return to the simple days of narrow banking. "The financial landscape has changed dramatically since then and banks have become bigger due to economic development, bringing with it economies of scale to customers and the broader economy." Furthermore, he believes the ongoing 'too big to fail' debate, which focuses on the return to narrow banking, overlooks the key problem of the interconnectivity between banks.

Dr. Banziger says new measures must be introduced to secure the interconnections across three key areas: payments and clearing, which are currently run through banks' balance sheets; the over-the-counter (OTC) derivatives markets, itself a huge web of interconnected contracts; and customer deposits. "We need to implement settlement systems that are removed from any risk of bankruptcy; efforts to provide a central counterparty for OTC clearing must be completed; and we must protect bank customers better through better deposit insurance schemes." He says that once measures have been introduced to secure these interconnections, a bank would be able to fail and the impact would be borne by shareholders and bondholders, not taxpayers. "Reforms like these are essential to restore confidence and will demonstrate that the question is not that banks are too big to fail, but rather that they are too interconnected to fail," says Dr. Banziger.

Transparency and standardisation will pave the way

These measures must also go hand-in-hand with improvements in transparency. Investors will not return to markets without reliable information on which they can base their investment decisions. This will mean pooling of information and greater disclosure on the underlying assets within structured products. "All relevant information should be publicly available so investors and supervisors can carry out their own independent risk assessment of each product," says Dr. Banziger. Additionally, the most important information must be made prominent and readily available to all investors.

Increased transparency needs to be coupled with higher standardisation to reduce complexity within the structured credit markets and improve liquidity. "More standardisation will improve the likelihood of market prices being available for products, particularly in difficult conditions," adds Dr. Banziger. A good example of the benefits of greater disclosure is the pfandbriefe (covered bonds) market in Germany. During the crisis this segment outperformed other securitisations due to better transparency, liquidity and standardisation.

Co-ordinated efforts required for liquidity pools

Liquidity has clearly been an issue in the recent crisis. This is an area that will require significant work, both internally within private institutions and broadly in terms of the co-ordination between central banks, financial supervisors and market participants. "Individual governments moved swiftly in the recent crisis to calm fears and moderate turbulence, but greater co-ordination internationally on liquidity and central bank instruments could have multiplied the impact," says Dr. Banziger.

Financial institutions must also re-evaluate the models they use to stress test for future crises. Banks have traditionally relied on stress-testing models to estimate the impact of external shocks on their profitability, liquidity and capital requirements, but the crisis has shown that the scenarios used were not extreme enough. As we have seen, if a bank finds itself with liquidity problems, this can quickly spiral into greater solvency problems. Asset prices are put under pressure, credit lines are withdrawn and it becomes more expensive to raise funds. This strengthens the case for banks to hold strategic liquidity reserves that cover on- and off-balance sheet funding for several months. This funding will enable firms to weather short-term volatility.

Learning from our forefathers

In conclusion, we are still in the midst of this crisis and much work needs to be done around transparency, liquidity and the financial system interconnectivity before we can restore trust and investor confidence. These are far-reaching measures that will require intellectual input from all corners and a great deal of collaboration and co-ordination to implement. "But just as our forefathers navigated their way out of the storm in the 1930s, we will steer a path that will return us to stability and growth," Dr. Banziger concludes.

The man-made financial crisis calls for drastic man-made solutions

The following reflects the personal views of Lars Rohde, CEO of Denmark's largest investor, the state pension fund ATP (Arbejdsmarkeds Tillægspension).

The man-made crisis

There are a number of factors that have contributed to the financial crisis. But the causes were all man-made, brought about by improper interventions by decision-makers such as:

– geopolitical events and decisions resulting in macroeconomic trends;
– deregulation by policy-makers;
– pro-cyclical monetary and fiscal policies;
– increasing risk appetite among investors;
– top managers ignored good corporate governance;
– excessive greed by top managers in financial institutions.

Since the fall of the Berlin Wall and in the wake of the IT bubble, the world has witnessed a period of constant growth and increasing asset prices. Gone was the memory of the last IT bubble as well as the Asian banking crisis. Most market participants shared the view that governments would always step in as a last resort if markets showed any signs of crisis ('The Greenspan Put').

The last decade has also been characterised by low inflation as well as low interest rates offered by central banks. This made it easy to get funding at favourable terms.

Financial globalisation increased tremendously, together with financial integration that eroded the traditional differences between, banks, insurance companies, pension funds etc. However, the regulatory framework as well as financial supervision did not keep pace with this rapid development. Regulation was fragmented and not comprehensive.

Prior to the crisis, financial market participants observed that risk indicators were reduced such as decreasing spreads in combination with increasing asset prices. This created large profit opportunities together with increased risk appetite, which eventually resulted in an asset bubble. In the search for higher returns, many investors bought structured products that they did not fully understand. As long as these structured products were rated by the rating agencies, investors bought them believing that liquidity would never constitute a problem. ATP did not invest heavily in structured products. As Lars Rohde explains "by slicing the cake one cannot make the cake larger or circumvent the laws of statistics".

Structured products, such as collateralised debt obligations (CDO), suffered from a lack of transparency, and Rohde adds that their extensive use was the main trigger for the financial crisis.

Solvency II – more actions by central banks

Rohde does not believe that the current solvency rules are sufficient to avoid similar financial crises in the future, as they are not capable of addressing systemic risk in an efficient manner. Larger capital buffers are needed but individual capital buffers cannot make the system as a whole safe. He also believes banks should be required to set aside more liquidity as all financial crises start with a funding problem.

"The crisis has taught us that there is a need to think outside the box," he says. "The present models should be re-evaluated, in particular with respect to how they capture extreme events.

Furthermore, central banks should play a more active role, as they do not have vested interests. Central banks should not only provide capital in order to avoid a collapse in the interbank market, but they should also seek to prevent asset bubbles in due time. "Moreover," Rohde suggests, "banks should pay for mandatory insurance against financial distress that should reflect each bank's total risk exposure."

The failure of good corporate governance

Rohde argues that owners become investors when listed companies are owned by a large number of anonymous shareholders who have no incentive to challenge incumbent management. This 'free rider' problem is embedded in dispersed ownership where there is a risk that management receives a call option written on the shareholders. It is optimal for small, diversified owners to vote with their feet, but on the other hand it is necessary that someone takes the responsibility to monitor management including their risk appetite.

Tougher legislation does not work as a substitute for concentrated ownership. Instead the US response was to promote a market for corporate control where different management teams struggled for control over listed corporations that were experiencing declining stock prices. However, Rohde finds that we cannot entirely rely on a market for corporate control. He emphasises the importance of a constructive dialogue with top management that is often practiced informally, in other words not at a general meeting. Even though large institutional investors may, to a certain extent, absorb the costs of active ownership, Rohde admits that sometimes it is frustrating that small investors expect the large shareholders to do the job. On the other hand, the active large investors may, over the longer term, get a clearer picture of management's skills and qualifications from such active dialogue.

The crisis has revealed that the governance structures proved insufficient in a number of large financial institutions. This means that one should change the existing governance setup, for example how top management is organised. What is needed is that board members have a contingency plan to handle large systemic risks. There is no need for board members to control all matters in detail, nor to set up very large and complex control mechanisms at board level. But the board should be able to challenge the CEO in a constructive manner, such as by asking critical questions about the firm's risk management based on worst case scenarios. Rohde notes that the level of sophistication in risk management increased tremendously during the nineties, but it did not capture the fact that the current financial crisis came from lack of liquidity.

Government bailout plans should be handled with care

State ownership should only be temporary in order to help financial institutions to recover from crises. Moreover, since the state has provided capital for financial institutions it is natural that the state also benefits financially when the situation improves.

However, Rohde stresses that the state should never interfere in the daily business matters of financial institutions or let its social agenda influence how state ownership is exercised. There is a need for a separation of banks into good and bad banks even though this might be difficult. Bad banks do not necessarily go bankrupt, but their balance sheets need to be cleaned up.

Governments as well as central banks have made various attempts to inject capital into the financial system. The aim is not only to save healthy banks, but also to stimulate the economy as a whole by helping non-financial firms with their liquidity problems as banks are reluctant to provide new loans to businesses. Rohde does not believe that governments can force banks to lend more capital to business, as this would not be practically possible to implement. This would in fact tend to force some banks into further liquidity problems. Moreover, banks are currently responding to the crisis by increasing the lending spread while hoping that the crisis will be over relatively soon, and this, in conclusion, calls for strong man-made initiatives.

Conclusion

These interviews show that there are several causes behind the current financial crisis. However, all three interviewees agree that insufficient regulation, unhealthy public and monetary policy as well as investor over-confidence constitute the main causes. These key representatives of large, leading financial institutions agree that there is a need to think out-of-the-box as the existing regulatory framework does not protect sufficiently against systemic risks. Trying to secure individual financial institutions does not secure the system as a whole and hence there is an urgent need to change the focus of regulation. Specifically, more focus should be directed towards ensuring sufficient liquidity during times of crisis and in which respect central banks should play a much more active role.

It is interesting to note that Lars Rohde pays relatively more attention to the importance of corporate governance compared to Tom Wilson and Hugo Banziger. All agree that good corporate governance should seek to facilitate proper risk management as well as securing well-balanced incentives. However, it is only natural that influential institutional investors put strong emphasis on good corporate governance as they are especially concerned with securing the interests of capital suppliers.

Moreover, all three agree that there should be much more focus on effective risk management at the top level of financial institutions. This does not entail a more direct involvement by the board in daily risk management processes. Instead, boards should become more concerned about determining the appropriate level of risk appetite in financial institutions as well as securing a sufficient liquidity buffer for times of crisis. Perhaps the most important lesson from the interviews is this: "Don't buy a product if you don't fully understand its risks" – a simple, but crucial lesson which will hopefully not be forgotten in the future.